Alexander Philipp

Delivery of siRNA with bioresponsive cationic-polymer based carriers

Alexander Philipp

Delivery of siRNA with bioresponsive cationic-polymer based carriers

Non-viral delivery of siRNA

Südwestdeutscher Verlag für Hochschulschriften

Impressum/Imprint (nur für Deutschland/ only for Germany)
Bibliografische Information der Deutschen Nationalbibliothek: Die Deutsche Nationalbibliothek verzeichnet diese Publikation in der Deutschen Nationalbibliografie; detaillierte bibliografische Daten sind im Internet über http://dnb.d-nb.de abrufbar.

Alle in diesem Buch genannten Marken und Produktnamen unterliegen warenzeichen-, markenoder patentrechtlichem Schutz bzw. sind Warenzeichen oder eingetragene Warenzeichen der jeweiligen Inhaber. Die Wiedergabe von Marken, Produktnamen, Gebrauchsnamen, Handelsnamen, Warenbezeichnungen u.s.w. in diesem Werk berechtigt auch ohne besondere Kennzeichnung nicht zu der Annahme, dass solche Namen im Sinne der Warenzeichen- und Markenschutzgesetzgebung als frei zu betrachten wären und daher von jedermann benutzt werden dürften.

Verlag: Südwestdeutscher Verlag für Hochschulschriften Aktiengesellschaft & Co. KG
Dudweiler Landstr. 99, 66123 Saarbrücken, Deutschland
Telefon +49 681 37 20 271-1, Telefax +49 681 37 20 271-0
Email: info@svh-verlag.de
Zugl.: Munich, Ludwig-Maximilians-University (LMU), Dissertation, 2010

Herstellung in Deutschland:
Schaltungsdienst Lange o.H.G., Berlin
Books on Demand GmbH, Norderstedt
Reha GmbH, Saarbrücken
Amazon Distribution GmbH, Leipzig
ISBN: 978-3-8381-1748-5

Imprint (only for USA, GB)
Bibliographic information published by the Deutsche Nationalbibliothek: The Deutsche Nationalbibliothek lists this publication in the Deutsche Nationalbibliografie; detailed bibliographic data are available in the Internet at http://dnb.d-nb.de.

Any brand names and product names mentioned in this book are subject to trademark, brand or patent protection and are trademarks or registered trademarks of their respective holders. The use of brand names, product names, common names, trade names, product descriptions etc. even without a particular marking in this works is in no way to be construed to mean that such names may be regarded as unrestricted in respect of trademark and brand protection legislation and could thus be used by anyone.

Publisher: Südwestdeutscher Verlag für Hochschulschriften Aktiengesellschaft & Co. KG
Dudweiler Landstr. 99, 66123 Saarbrücken, Germany
Phone +49 681 37 20 271-1, Fax +49 681 37 20 271-0
Email: info@svh-verlag.de

Printed in the U.S.A.
Printed in the U.K. by (see last page)
ISBN: 978-3-8381-1748-5

Copyright © 2010 by the author and Südwestdeutscher Verlag für Hochschulschriften Aktiengesellschaft & Co. KG and licensors
All rights reserved. Saarbrücken 2010

Table of Contents

1 INTRODUCTION 1

1.1 Nucleic acid based therapy: applications in tumor therapy 1

1.2 Non-viral carrier systems for nucleic acid based therapy 4

1.3 Extra- and intracellular barriers for nucleic acid delivery 6

1.4 Design of bioresponsive polymers with virus-like functionalities 9
  - 1.4.1 Shielding functionality 10
  - 1.4.2 Targeting functionality 11
  - 1.4.3 Endosomal release functionality 12
  - 1.4.4 pH responsive and redox sensitive systems 13

1.5 Aims of the thesis 15

2 MATERIALS AND METHODS 17

2.1 Chemicals, polymers and other reagents 17

2.2 Additional novel polymer conjugates 19
  - 2.2.1 Synthesis of succinic anhydride (Suc) modified OEI-HD-1 19
  - 2.2.2 Synthesis of PEG modified OEI-HD-1 19
  - 2.2.3 Synthesis of 3-(2-pyridyldithio)-propionate modified PEG-OEI-HD-1 .. 20
  - 2.2.4 Synthesis of DMMAn-Mel modified PEG-OEI-HD-1 21

2.3 Biophysical characterization 21
  - 2.3.1 siRNA binding ability 21
  - 2.3.2 Polyplex formation 22
  - 2.3.3 Agarose gel retardation 22
  - 2.3.4 Polyplex stability against sodium chloride 22
  - 2.3.5 Particle size and zeta-potential measurement 22
  - 2.3.6 Transmission electron microscopy 23

2.4 Biological characterization 23
  - 2.4.1 Cell culture 23
  - 2.4.2 Luciferase reporter gene silencing 24
  - 2.4.3 Metabolic activity of cells after polymer treatment 24
  - 2.4.4 Hemolytic activity of polymers 24
  - 2.4.5 Reverse Transcriptase quantitative real-time PCR (RT-qPCR) 25

Table of Contents

  2.4.5.1 RNA isolation and cDNA synthesis 25
  2.4.5.2 Quantitative real-time PCR... 25

2.5 Statistics.. 26

3 RESULTS 27

3.1 Modified PEIs with reduced toxicity as efficient siRNA carriers 27
 3.1.1 Design of PEI conjugates with reduced charge density... 27
 3.1.2 siRNA binding and complexation ability 29
 3.1.3 Influence of conjugates on cytotoxicity 31
 3.1.4 siRNA delivery efficiency: structure-activity relationship... 31
 3.1.5 Study on mechanism of the highly effective siRNA carrier PEI-Suc-10% 34
 3.1.6 Polyplex stability against sodium chloride induced dissociation 35

3.2 Biodegradable OEI conjugates for siRNA delivery 37
 3.2.1 Polymers based on oligomerized OEIs (OEI-HD-1) 37
  3.2.1.1 Design of OEI-HD-1.. 37
  3.2.1.2 siRNA complexation ability 38
  3.2.1.3 siRNA delivery efficiency and toxicity of OEI-HD-1.. 39
  3.2.1.4 Transferrin receptor targeting of siRNA polyplexes 42
  3.2.1.5 Succinylated OEI-HD-1 for improved effective window..................................... 43
 3.2.2 Pseudodendritic oligoamines 46
  3.2.2.1 Design of OEI core based conjugates... 46
  3.2.2.2 siRNA delivery efficiency: structure-activity relationship 47
 3.2.3 Hydrophobically modified OEIs 50
  3.2.3.1 Design of modified OEI conjugates 50
  3.2.3.2 siRNA complexation ability 52
  3.2.3.3 Colloidal stability of polyplex particles 53
  3.2.3.4 Influence of conjugates on cytotoxicity 55
  3.2.3.5 siRNA delivery efficiency: structure-activity relationship 56
  3.2.3.6 Lytic activity of conjugates... 59
  3.2.3.7 Co-formulation with helper polymers and lipids for improved siRNA delivery ... 60

3.3 Bioresponsive endosomolytic conjugates for siRNA delivery 62
 3.3.1 DMMAn-Mel modified conjugates for siRNA delivery 62
  3.3.1.1 Design of endosomolytic conjugates 62
  3.3.1.2 siRNA binding ability 64
  3.3.1.3 siRNA delivery efficiency and toxicity of conjugates 65
 3.3.1.3.1 PEG-PLL conjugates 65
 3.3.1.3.2 PEG-PEI conjugates 67
 3.3.1.3.3 PEG-OEI-HD-1 conjugates 68

3.3.2 Covalently attached siRNA-polymer conjugates for improved siRNA delivery 70
    3.3.2.3 Design of dynamic functionalized siRNA carriers 70
    3.3.2.4 Particle size determination of conjugates and polyplexes 72
    3.3.2.5 siRNA delivery efficiency and toxicity of PEG-PLL-DMMAn-Mel-siRNA conjugates .. 73
    3.3.2.6 pH triggered lytic activity of conjugates 75
    3.3.2.7 Glutathione induced release of siRNA 76

4 DISCUSSION 77

4.3 Evaluation of modified PEIs with reduced toxicity as efficient siRNA carriers 77
    4.3.1 Improved biological properties of modified PEIs 78
    4.3.2 Structural requirements for efficient siRNA delivery 79

4.4 Evaluation of biodegradable OEI conjugates for siRNA delivery 80
    4.4.1 Oligomerized OEIs: targeting for optimized virus-like siRNA delivery 81
    4.4.2 Pseudodendritic oligoamines with high potential for siRNA delivery 82
    4.4.3 Hydrophobically modified OEIs: structural influence on biological activity 83

4.5 Evaluation of bioresponsive endosomolytic conjugates for siRNA delivery 85
    4.5.1 DMMAn-Mel modification for enhanced siRNA delivery efficiency 86
    4.5.2 Dynamic siRNA-polymer conjugates for programmed delivery 87

5 SUMMARY 90

6 ABBREVIATIONS 93

7 REFERENCES 96

8 ACKNOWLEDGEMENTS 117

1 INTRODUCTION

1.1 Nucleic acid based therapy: applications in tumor therapy

Nucleic acid based therapy offers a promising strategy in the treatment of cancer or many other genetic (e.g. cystic fibrosis, severe combined immunodeficiency) and acquired (e.g. infectious, neuropathological) diseases by delivering therapeutic nucleic acids into patients.

While traditional gene therapy utilizes DNA to correct a genetic defect by inserting functional genes into an organism in order to replace defective ones, strategies in cancer therapy range from inserting tumor suppressor genes to immunotherapy[1]. Gene therapy was already applied in various clinical studies[2-4] using genes, for example, encoding for antigens, cytokines, tumor suppressors or different growth factors and receptors.

Table 1 shows the different types of therapeutic nucleic acids which have been delivered into the target cell by non-viral transfer systems.

These various types of nucleic acids achieve different effects at the molecular level. In non-viral plasmid DNA (pDNA), for example, is mainly used for intra-nuclear delivery to replace or to substitute a specific genetic function in the target cell resulting in a "gain of gene function". In contrast, "loss of gene function" is often mediated by intra-cytoplasmatic delivery of synthetic asRNA or siRNA reducing the expression of endogenous genes in a sequence-specific manner, which can be used for silencing of pathogenic target genes or inducing specific antitumoral effects[5-7].

Novel strategies of exploiting the antisense mechanism include triggered exon-skipping for partial repair of defective genes[8] and targeting miRNAs with complementary oligoribonucleotides (anti-miRNA)[9-10] in order to improve gene expression.

Among these, siRNA mediated gene silencing has attracted considerable research interest, since the discovery of RNA interference (RNAi) by Fire et al in 1998[11]. Significant efforts and resources have been currently invested in exploiting the therapeutic potential of siRNA for the treatment of human diseases such as cancer[6-7,12-17].

1 INTRODUCTION

Nucleic acid	Description
pDNA	plasmid DNA containing gene cassettes for expression of proteins, antisense RNAs or short hairpin silencing RNAs
mRNA	delivery of mRNA into the cytosol for protein expression, for example antigens in dendritic cells to stimulate immunity[18-19]
asRNA	single stranded antisense RNA, binds to complementary mRNA strands, inhibits gene expression[20-21]
siRNA	single stranded small interfering RNA, binds to complementary mRNA strands followed by catalytical cleavage, inhibits gene expression[11,22-23]
miRNA	single-stranded micro RNA, binds to partial complementary mRNA strands, inhibits gene expression
polyIC	poly-inosine-cytosine double-stranded RNA, interacts with endosomal toll-like receptor 3 and cytosolic mda-5 receptor, triggers apoptosis and interferon response[24]
decoy DNA	oligodeoxynucleotide decoy, binds to transcription factors via consensus sequences as in target genes, inhibits transcription factor functions and gene expression[25]
DNA/RNA aptamers	DNA or RNA oligonucleotides, bind to a specific target molecules (e.g. nucleic acids, proteins), inhibit target molecule functions[26]

Table 1: Different types of therapeutic nucleic acids delivered by non-viral (physical or chemical) methods.

The RNA interference process makes use of double stranded RNAs for sequence specific gene silencing. The introduction of exogenous long dsRNA (> 30 nucleotides) into cells was found to inhibit cellular protein expression, but it additional provokes innate immune response by interferon activation causing apoptosis. Tuschl and colleagues demonstrated that target gene-specific RNA interference without significant side effects can be mediated by application of small synthetic 21-23 nucleotide based RNA duplexes[22-23,27]. The mechanism of RNA interference is presented in Figure 1, an endogenous process, which employs small interfering RNAs (siRNAs) to suppress target-specific gene expression by mRNA degradation.

1 INTRODUCTION

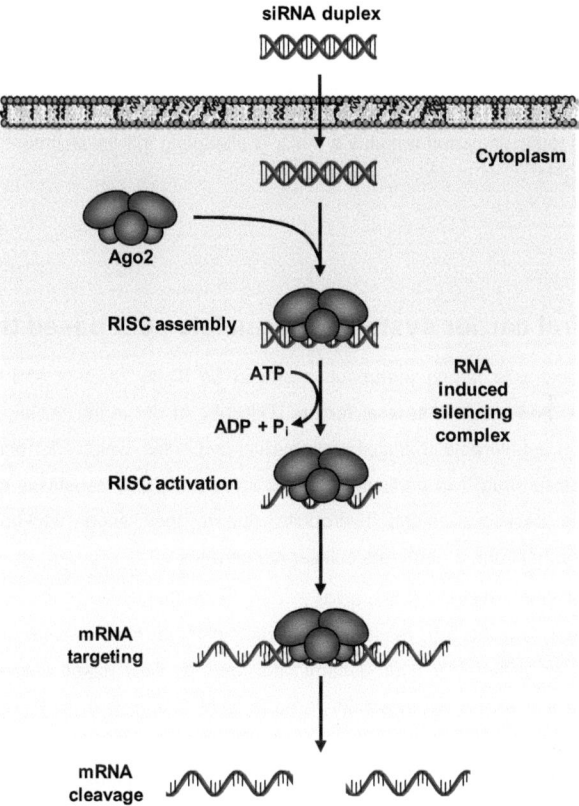

Figure 1: The mechanism of RNA interference. After entering the cytoplasm, siRNA duplexes are incorporated into the RNA induced silencing complex (RISC) resulting in cleavage of the sense strand of siRNA by the enzyme argonaute 2 (Ago2). The activated RISC complex cleaves target mRNAs with complementary domains due to its endonuclease activity resulting in sequence specific gene silencing.

Within the cytoplasm of cells, siRNA gets incorporated into a ribonucleotide protein complex called RNA induced silencing complex (RISC). After activation, siRNA becomes unwound by the enzyme argonaute 2 and the sense strand is cut off[28-30]. The antisense strand remains incorporated and triggers cleavage of mRNAs with complementary domains due to the catalytic nature of RISC. Cleavage of the mRNA leads to reduction of translation and, thus, in target specific gene silencing. With increasing knowledge on the molecular mechanisms of endogenous RNA interference, synthetic siRNAs promise great potential as a new class of therapeutic nucleic acids in the treatment of various forms of cancer and a number of other diseases, due to their ability to silence gene expression in a sequence specific manner[31-33].

Although, numerous siRNA and related nucleic acid formulations have been already applied in clinical trials[4,34-35] and have also shown very encouraging anticancer effects in vivo, such as inhibition of neoangiogenesis[36-38], induction of apoptosis[24,39-40] or reduction of tumor cell proliferation[41-43], the lack of safe and efficient delivery systems still limits the full therapeutic potential of this technology and remains a major challenge in the development of nucleic acid based therapies[44-47].

1.2 Non-viral carrier systems for nucleic acid based therapy

Generally, nucleic acid based therapeutics should be highly efficient and well tolerated, which depends, however, on several factors. Difficulty in delivering nucleic acids already results from their unfavorable chemical and physical properties, which are not consistent with that of a successful drug[48]. In particular, nucleic acids are highly negatively charged due to their phosphate backbone, highly hydrophilic due to their sugar backbone and large macromolecules that cannot permeate cellular membranes.

Thus, direct delivery of naked nucleic acids in vivo, i.e. in the absence of a carrier molecule, can be only rarely applied with reasonable efficiency[16,49], such as in case of intramuscular injection of naked pDNA[50] or hydrodynamic delivery[51-53]. Also naked siRNA formulations have only been successful, when administered to local tissues, e.g. by direct injection into the eye for the treatment of age related macula degeneration[54-55], which has already been applied in humans in clinical trials[56]. However, systemic applications of nucleic acids, which are required for the broad range of indications such as disseminated cancer, involve further problems like undesired interactions with blood components, susceptibility of rapid enzymatic degradation by serum nucleases and clearance from the bloodstream (mainly by Kupffer cells in the liver) resulting in short half life times of a few minutes[57-60].

In order to succeed in clinical application of nucleic acid based therapy, safe and efficient carrier systems have to be developed, which are able to stabilize nucleic acids in the extracellular environment and effectively deliver them into the target site. For this purpose, various viral and non-viral delivery strategies have been discovered.

Viral vectors, especially retroviral and adenoviral vectors, are the most commonly used nucleic acid delivery vehicles in clinical trials due to their high delivery efficiency[4,61-62] and their facility for in vivo tissue-specific replication, which is very useful for clinical applications like in cancer. However, their broad applicability may be limited in terms of safety concerns[63-64] regarding their immunogenicity[65] and carcinogenicity, which is caused by mutagenesis resulting from gene insertion into the host genome[66-67] and their additional high production costs and low capacity to incorporate therapeutic nucleic acids.

1 INTRODUCTION

For this reason, non-viral vectors have been investigated as alternatives with useful characteristics, such as enhanced biosafety and pharmaceutical advantages meaning simple synthesis and large-scale production. However, synthetic vectors show in general far less efficiency compared to their viral counterparts after in vivo application[68-71].

Cationic lipid[72-75] and cationic polymer[76-79] based systems are the major types of non-viral carrier systems, which package and condense the negatively charged nucleic acids into particles of virus like dimensions protecting them from degradation[80]. Carrier molecules interact with nucleic acids in a reversible, non-damaging manner which is in most cases provided by electrostatic interactions between the positively charged groups on the carrier molecule and the negatively charged phosphate groups on the nucleic acid. Also covalent attachment of nucleic acids to carrier molecules has been exploited[81-85]. Cell entry mainly occurs via adhesion of the positively charged carrier systems to the negatively charged transmembrane heparin proteoglycans followed by endocytosis[86-87].

Lipoplexes[88] are based on non-covalent complex formation of cationic lipids, such as DOTAP or DOTMA representing amphiphilic molecules with a hydrophobic tail and a hydrophilic cationic head group, with negatively charged nucleic acids[74,89-91]. Lipid formation takes place by spontaneous aggregation of cationic surfactants with hydrophobic moieties (long alkyl chains) forming hydrophobic interiors and positively charged polar head groups, which results in a polycationic surface. After cellular uptake, lipoplexes are able to lyse endocytotic vesicles caused by provoking membrane pertubation[92-96] and subsequently destabilized lipoplexes break down resulting in efficient release of the nucleic acids into the cytoplasm[97-98]. However, a major disadvantage of cationic lipid based carrier systems is that they are relatively unstable in physiological (i.e. serum containing) environment, which strongly restricts their use for extensive in vivo applications.

Polyplexes[88] are based on non-covalent complex formation of cationic polymers, such as PLL or PEI, with negatively charged nucleic acids[70,99]. Poly-L-lysine (PLL) was one of the first polycations used for polyplex formation[71], which represents a biodegradable polymer due to its natural amino acid backbone. At physiological pH the primary amino groups of PLL are positively charged and interact electrostatically with the negatively charged nucleic acids to complex them into nanoparticles. Although the cellular uptake of the polyplexes is effective, subsequent escape from endosomes into the cytoplasm presents a major bottleneck for this formulation. Over the past decade polyethylenimine (PEI), first introduced by Behr and colleagues[100], has become one of the most commonly used polycations for nucleic acid delivery and even has been considered as the golden standard in many in vitro and in vivo applications[24,101-104]. PEI is a non-degradable polymer with considerable buffering capacity below the physiological pH promoting endosomal escape to a certain degree due to the so-called osmotic burst or "proton sponge" effect. Osmotic swelling in combination with direct interaction of the polycation with the inner endosomal membrane cause local endosomal membrane disruption leading to a release of the nucleic acid carrier into the cytoplasm[105-108]. Polyethylenimine based polymers can be synthesized with different molecular weights in a linear or branched structure or can undergo functionalization by group addition or substitution, which strongly influences nucleic acid delivery efficiency and toxicity[109-111]. However, the major drawback of the formulation seems to be pronounced toxicity[112-114], both in vitro and in vivo, due to the huge amount of positive surface charges and a variety of unspecific interactions with the biological environmen[115-118]. Moreover, insufficient metabolization and elimination due to the lack of biodegradability finally result in unintentional accumulation in cells and excretion organs, such as liver, which further limits the applicability in vivo for repeated systemic administration. Thus, degradability of polycations has become a crucial factor for practicability and safety of in vivo applications, which was taken into account for the development of novel biodegradable polycations for nucleic acid delivery[45,111,119-136].

Besides synthetic polymers, various biopolymers have been additionally exploited as non-viral delivery vehicles, which are also able to bind or encapsulate nucleic acids into nano-sized complexes, such as chitosan[137-141], gelatine[142-145], atelocollagen[146-147] or other nanoparticles[148-151].

1.3 Extra- and intracellular barriers for nucleic acid delivery

Nucleic acid carrier system should protect and deliver nucleic acids efficiently and exclusively to the site specific target cells. However, on their in vivo delivery route after systemic application, carriers are faced with numerous extracellular and intracellular barriers[152] as shown in Figure 2.

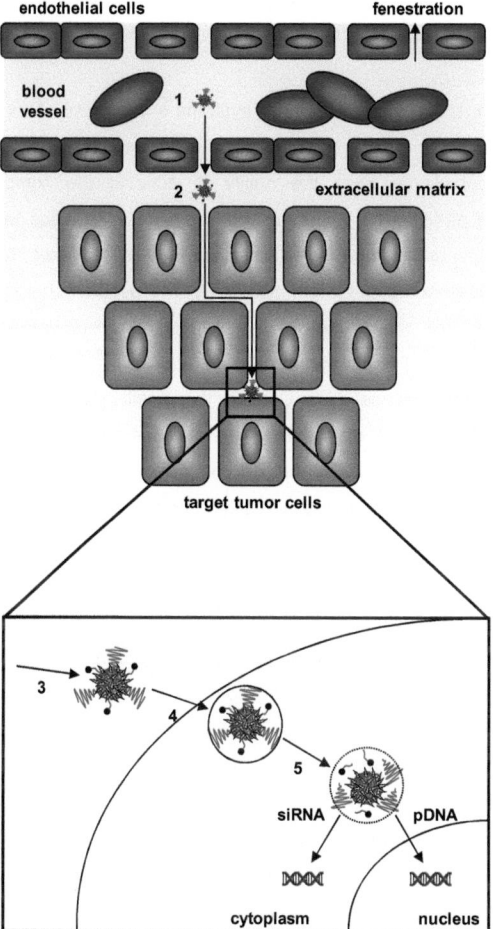

Figure 2: Physiological barriers for systemic delivery of nucleic acids. After systemic application, delivery vehicles have to (1) avoid undesired interactions with blood components, aggregation, degradation or complement activation in the bloodstream, (2) cross the leaky tumor vascular endothelial barrier into the interstitial space, (3) diffuse through the extracellular matrix towards the target tumor tissue, (4) be internalized specifically into the target cells, (5) escape from the endosome and disassemble and release the nucleic acid payload within the proper intracellular compartment (nucleus for pDNA, cytoplasm for siRNA or mRNA).

Upon systemic application, complexes have to survive in the bloodstream as they are surrounded by a variety of blood compounds, such as erythrocytes, salts, lipids, carbohydrates, serum proteins or degradative enzymes. This may also influence the composition of the complexes causing aggregation, dissociation or degradation as well as the bioavailability due to the fact that other charged molecules can disrupt such complexes before they reach the target cell. Thus, even when reaching the target cells, the complexes may no longer exhibit the physical properties necessary for efficient nucleic acid delivery into the cells.

For delivery to distant target sites the carrier systems should also show elongated plasma circulation times. However, as a result of introducing foreign molecules into the body, positively charged complexes have the ability to activate the complement system[115]. Opsonization of such particles by the complement protein C3b leads to the initiation of a cascade of events presumably resulting in fast clearance of comlexes due to phagocytosis by cells of the reticulo-endothelial-system. Moreover, the positive charge of complexes not only mediates target cell attachment and internalization, but also causes unspecific interactions with negatively charged membranes of blood components, vascular endothelial cells or other non-target tissues[115-118]. Coating the positive charges of lipoplexes and polyplexes with other hydrophilic macromolecules, such as polyethylene glycol (PEG), was found to avoid unspecific interactions with blood components and recognition by the immune system, which finally resulted in prolonged circulation times[116,153-156].

In order to reach the target cells, complexes have to extravasate across the leaky tumor vasculature into the interstitial space, where interactions with the extracellular matrix have to be avoided. The extracellular matrix comprises different combinations of collagens, proteoglycans, hyaluronic acid, fibronectin and other glycoproteins, i.e. components that could act as further hurdle by binding to the complexes. It was found that lipoplexes and polyplexes resulting in a net positive charge, could also interact with the extracellular matrix causing dissociation or aggregation, which consequently negatively affects the delivery efficiency[157-158]. Probably complex disassembly is more likely a problem in the case of siRNA compared to pDNA, as the far larger number of negative charges offered in pDNA stabilizes the interelectrolyte complex[159-160]. However, to reach the target site of action the nucleic acid has to stay associated with its carrier during the complete extracellular delivery process. Hence, either stabilization of comlexes using lateral reversible crosslinking strategies after complex formation[131,161-163] or covalent attachment of the nucleic acid to the carrier system[81-85] have been found to be a promising tool to overcome the undesired instability of complexes in the presence of serum proteins or the extracellular matrix.

At the target site, complexes have to bind specifically to the target cells, which is of crucial impact, as interaction with non-target cells could trigger undesired and potentially toxic side effects. Selective targeting to cell type specific tissues can be achieved by incorporation of targeting ligands into the carrier systems recognizing cell type specific receptors expressed on the cell surfaces resulting in cellular uptake via receptor mediated endocytosis[164-165].

After successful internalization into the target cells, the complexes are located in intracellular endosomal vesicles. Acidification of endosomal vesicles prepares endosomes to fusion with lysosomes containing digestive enzymes and nucleases for degradation of the endosomal content. Thus, the release of the complexes out of the endosomes into the cytoplasm represents a major challenge to achieve effective nucleic acid delivery, as endosomal entrapment is associated with degradation of the nucleic acids upon endosomal acidification.

Following endosomal escape, pDNA complexes still have to enter the nucleus after cytoplasmic trafficking in order to reach the transcriptional / translational machinery[166]. Hence, nuclear import of pDNA followed by carrier unpacking represents another major hurdle which has to be overcome. In contrast, for siRNA complexes the cytoplasm is the target site of action and effective dissociation of siRNA from the carrier systems in the cytoplasm is required for assembly of the RISC complex.

Once the vectors have delivered their therapeutic nucleic, they should be easily metabolized and being eliminated in order to avoid accumulation in organs which results in undesired long term toxicity. Particle size represents another general critical factor for drug targeting[167], as several hundred nm large complexes are not able to penetrate endothelial barriers[168] or extravasate from the leaky tumor vasculature into the interstitial space due to size restrictions or even may trigger acute toxicity after systemic application.

Altogether, nucleic acid carrier systems have to meet many requirements for successful delivery of nucleic acids to their target sites as a series of intracellular and extracellular barriers have to be overcome before the delivered nucleic acids can achieve their full therapeutic potential.

1.4 Design of bioresponsive polymers with virus-like functionalities

For successful nucleic acid based therapy, the development of appropriate carrier systems for nucleic acid delivery is a major challenge, since all of these aspects have to be taken into consideration.

Thus, novel biodegradable carriers are needed, which exhibit improved efficiency, less toxicity and a better biocompatibility[45,111,119-136]. Moreover, in order to meet all the requirements for overcoming all the biological delivery steps, nucleic acid carriers will have to mediate different functions upon their delivery pathway, meaning that they have to be less static and should respond more dynamically to the cellular microenvironment they are exposed to. These requested functions can be pre-programmed into a carrier system by introduction of various functional domains resulting in a so-called "synthetic virus"[70,169-171] which faces virus-like functionalities as shown in Figure 3.

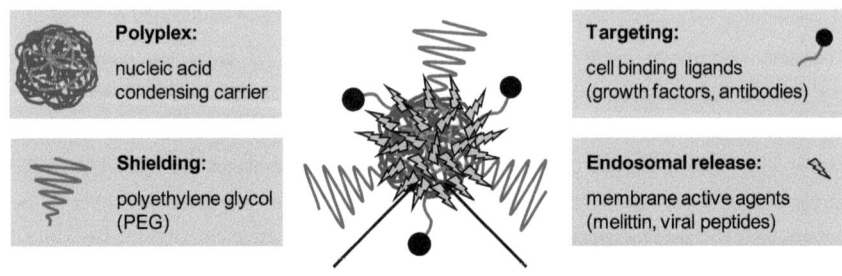

molecular sensors (chemical, physical)

Figure 3: Assembly of polyplexes. Besides shielding, targeting and endosomal release functionalities, nucleic acid delivery systems are additionally equipped with molecular sensors for programmed nucleic acid delivery. The acidic or reducible cellular microenvironment triggers cleavage of implemented bioresponsive elements in the nanoparticles enhancing the escape from intracellular vesicles into the cytoplasm where further disassembly might be triggered for efficient release of the payload from the carrier system.

Such synthetic virus-like carrier systems are more dynamic in their characteristics, like natural viruses, to be most effective at the different steps of extracellular and intracellular delivery. Mimicking the efficient, dynamic delivery process of viral infection[70,169,172], "artificial viruses" have been provided with a combination of certain functional attributes in order to protect the nucleic acid from degradation during circulation in the blood stream[154,173-174] and shield against unspecific interactions, facilitate targeting to specific cell surface receptors followed by cellular uptake at the target site and trigger efficient release of nucleic acids into the cytoplasm.

1.4.1 Shielding functionality

In order to prevent undesired interactions with blood components during the delivery process carrier systems were functionalized with hydrophilic molecules, such as polyethylene glycol[116,153,155-156], poly(N-(2-hydroxypropyl)methacrylamide) (pHPMA)[175-176] or poloxamer[177], resulting in reduced susceptibility to aggregation with serum proteins and phagocytosis by cells of the reticulo-endothelial-system. Furthermore, such sterically stabilized particles with neutral surface charge exhibit prolonged circulation times in the blood and enable passive accumulation in tumor tissue due to the enhanced permeability and retention (EPR) effect[178]. This "passive tumor targeting" of nanoparticles, relies on a leaky tumor vasculature combined with inadequate lymphatic drainage.

Different PEGylation strategies have been proven to be successful, e.g. direct attachment of PEG to the carrier system[173,179-180] or to the nucleic acid[151,181] prior to complexation (pre-grafting) as well as coupling of PEG to the polyplex surface after complex formation (post-grafting)[182-183]. The pre-PEGylation strategy enables the incorporation of a defined amount of hydrophilic polymer to the polycation, whereby the post-PEGylation strategy does not negatively influence the complexation process.

PEGylation, however, also negatively affects the cellular uptake process and escape from intracellular vesicles due to the fact that shielding with PEG molecules reduces the positive surface charge of polyplexes, which is an important factor for interaction with cellular membranes and in consequence for cell entry and endosomal release. This loss of efficiency can be partly compensated by introducing of targeting ligands or pH responsive and redox sensitive systems into the polyplexes.

1.4.2 Targeting functionality

Efficient delivery of nucleic acids can only succeed in therapy if properly directed towards the target site. Specific tissue targeting can be achieved by incorporation of targeting ligands into nucleic acid delivery systems recognizing cell type specific receptors on cell surfaces in order to promote cellular uptake via receptor-mediated endocytosis. The first receptor targeted polyplexes and their in vivo use was described already twenty years ago[184-185], as meanwhile many different tageting ligands have been evaluated, notably for their specific attachment to tumor cell surfaces[153,164-165,169,186-190].

Transferrin (Tf), a 79kDa iron transporting serum glycoprotein, has been widely studied as a ligand for tumor targeted delivery over many years[35,40,173,191-193]. Due to their high metabolism, cancer cells overexpress Tf receptors on their cell surface and hence can be effectively targeted with Tf formulated conjugates in vitro and in vivo[194-199]. Moreover, Tf as part of a carrier system also exerts an additional shielding function based on its relatively large size and slightly negative charge preventing unspecific interactions with blood components[194].

The epidermal growth factor (EGF) has also attracted much attention as a possible targeting ligand, as the EGF receptor is strongly overexpressed in many types of cancer[200-201]. EGF is a relatively small protein with a molecular weight of approximately 6kDa. Coupling to PEI, e.g via disulfide bonds[200] or a PEG spacer[153,179], as well as to other polycations resulted in greatly enhanced specifity and efficiency of nucleic acid delivery[24,202].

The arginine-glycine-asparagine (RGD) motif of fibronectin has also been studied for tumor targeted delivery of nucleic acids based on its capability to bind to the integrins that are expressed on the activated endothelial cells in tumor vasculature[203-206].

Folic acid, a vitamin which is necessary for the synthesis of purines and pyrimidines, is another strategy to achieve site specific tumor targeting. Attachment of folic acid to various nucleic acid delivery systems has been investigated for target specific delivery into tumors[207-209].

Besides natural ligands also anti-receptor antibodies, antibody fragments or completely synthetic ligands can be used for targeted delivery of nucleic acids[210-216].

1.4.3 Endosomal release functionality

Capture of non-viral carrier systems in endocytosed intracellular vesicles after cell entry is a major barrier for nucleic acid delivery. Ways of overcoming endosomal entrapment in order to avoid endo-lysosomal degradation are, for example, making use of the "proton sponge" effect of some polycations or the incorporation of lytic moieties into the carrierr system resulting in membrane disruption and, thus, endosomal escape into the cytoplasm.

Polycations, such as PEI possess considerable buffering capacity below the physiological pH promoting endosomal escape to a certain degree due to the so-called osmotic burst or "proton sponge" effect. This hypothesis is based on the chemical structure of PEI, which is only partially protonated at physiological pH, as approximately only every third nitrogen is positively charged. Hence, during the endosomal acidification process, protonation of the remaining secondary and tertiary amines acts like a "proton sponge" which is responsible for an increased osmotic pressure in the vesicles. Osmotic swelling in combination with direct interaction of the polycation with the inner endosomal membrane-cause local endosomal membrane rupture leading to a release of the nucleic acid carrier into the cytoplasm[105-108]. However, endosomal escape represents a major bottleneck when only small amounts of PEI are accumulated per endosome[217].

In addition, histidine or other imidazole containing polycations, which can also become protonated in the acidic endosomal environment, represent another approach to generate nucleic acid carrier systems capable of osmotic burst mediated endosomal release[218-223].

But not only polymers with buffering capacity can induce efficient release into the cytoplasm, also several cationic lipids possess endosomal escape properties. Lipid based formulations are able to lyse endocytotic vesicles by irreversibly absorbing lipids that are spontaneously released from the structurally dynamic endosomal membrane provoking membrane pertubation[92-96].

Another versatile method is the incorporation of membrane disrupting agents[224-226] into the carrier system mimicking the natural endosomal escape mechanisms achieved by cell invading organisms such as bacteria or viruses. Lytic artificial peptide sequences or natural sequences derived from Listeria monocytogenes[227], adenovirus[228-229], influenza virus[224,230] rhinovirus[231] or other virus-derived peptides[232-233] have been successfully applied for that purpose. Synthetic artificial amphipatic peptides that mimic natural lytic peptides were also designed. GALA (repeating units of glutamic acid-alanine-leucine-alanine) and KALA (repeating units of lysine-alanine-leucine-alanine) represent such basic amphiphathic peptides. Protonation at acidic pH triggers conformational changes from a random coil to an amphipatic alpha-helix exposing hydrophobic domains, which can interact with lipid bilayers and lead to membrane rupture[226,234-237]. Furthermore, melittin, a cationic lytic peptide derived from bee venom, has also been shown to strongly enhance the delivery efficiency of lipoplexes and polyplexes in vitro and in vivo[238-241].

1.4.4 pH responsive and redox sensitive systems

Closer observation of the delivery process reveals that nucleic acid carrier systems not only have to exhibit different delivery functions, but also have to meet different requirements at different time. For example, carrier systems should strongly bind and protect nucleic acids in the extracellular environment, but efficiently release it into the cytoplasm after cellular uptake. Shielding with PEG should prevent undesired interactions with proteins and cell membranes during systemic circulation, but within the endosomes the intracellular release is hindered as increased membrane interaction facilities are required to ensure endosomal escape. Although the endosomal membrane has to be destabilized for efficient delivery, pronounced membrane lytic activity, e.g. based on membrane active agents, is unfavorable outside of endosomes and, thus, has to be miimized in the extracellular environment.

As the described different delivery functions are required at different time points of the extracellular and intracellular delivery process, formulations can be pre-programmed to undergo dynamic changes and alter their characteristics during the delivery process like natural viruses, which sense their environment and respond to the biological surrounding in a dynamic manner[70,171,242]. The requested changes can be programmed into the synthetic carrier systems ("artificial viruses") by introduction of bioresponsive elements ensuring that the particular functions are only active during the phases of the delivery process were they are required. Thus, molecular sensors, such as hydrolytic cleavable bonds or reducible disulfide bonds have been utilized, which are able to respond to biological stimuli triggered by small changes in the cellular microenvironment[172]. As biological triggers, for example, differences in the pH of biological compartments[85,243-247] or different redox potentials inside and outside the cells[248] have been exploited.

While PEG shielding is important for the first steps of nucleic acid delivery, at later steps the shielding coat is no longer required and becomes even counter-productive, as it may prevent efficient nucleic acid release from the endosomes due to hindered destabilization of vesicular membranes[217]. For triggered deshielding, endosomal acidification can be exploited as biological stimulus, since the extracellular and intracellular pH values are physiological neutral. pH-labile chemical bonds such as acetals[134,243,249-251], hydrazones[245,252-255], orthoesters[91,256-258], thiopropionate linkers[259-261], dialkylmaleic acid monoamides[84,111] or vinyl ethers[123,262] have been utilized for removal of the PEG shield in the acidic endosomal environment, strongly improving the nucleic acid delivery process[174,242,263-264].

Moreover, lytic activity of membrane active peptides, such as melittin, is an unfavorable effect in the extracellular environment regarding toxicity of the delivery vehicles, as the membrane destabilizing activity is not only focused to intracellular vesicles. In an analogous fashion, masking the lytic activity of the peptide melittin with pH reversible chemical bonds, e.g. coupling the primary amines of melittin lysine residues to dimethylmaleic anhydride, strongly reduces the lytic activity of the peptide at neutral pH, which is recovered following endosomal acidification due to cleavage of the maleamate protecting groups[85,247,265-266]. Other recently developed dynamic nucleic acid delivery systems including the activation of endosomolytic properties[236,267-268] highlight the superiority and increasing impact of dynamic delivery systems in comparison with their static counterparts.

Exploiting the differences in the redox potential between oxidizing extracellular environment and reducing intracellular compartment of cells offers another possibility to alter the properties of synthetic carrier systems during the delivery process. Disulfide bonds, for example, which are stable in blood circulation, enable site specific cleavage in reducing intracellular environment upon cell entry[241,269]. Hence, covalent coupling of nucleic acids to the carrier via reductive cleavable disulfide bonds can be utilized to mediate extracellular stability, i.e. preventing dissociation of the delivery vehicle, but intracellular release of the nucleic acids following cleavage of disulfide linkages[84-85,222,270].

Besides bioresponsive deshielding or triggered disassembly of non-viral delivery vehicles, the differences in pH and redox gradient can also be used for development of biodegradable carrier systems in order to reduce their inherent toxicity. Bioreversible crosslinking of low molecular weight polymers into larger molecules, either by different hydrolytic cleavable or reducible disulfide linkages for triggered degradation into smaller fragments, represents an encouraging strategy, which resulted in improved delivery efficiency associated with less toxicity and better biocompatibility[45,111,119-136].

The development into pre-programmed bioresponsive systems, containing targeting ligands, shielding domains and membrane active moieties represents an important step in the field of nucleic therapeutics[79].

1.5 Aims of the thesis

The lack of appropriate delivery systems limits the novel and encouraging therapeutic and clinical potential of siRNA for the treatment of various diseases, such as cancer. Thus, the major focus of the current thesis was the discovery and optimization of novel polymers as highly effective and biocompatible siRNA delivery systems.

Polyethylenimine has proved to be one of the most widely used polycations for nucleic acid delivery, which however displays significant toxicity and - in case of siRNA delivery - only modest activity. The first aim of the thesis was to generate less toxic PEI derivates for siRNA formulation with improved biological properties. For that purpose, several modifications of PEI 25 had to be carried out to reduce the highly positive surface charge of the polycation, to optimize the therapeutic window of the formulation for siRNA delivery.

The second aim of the thesis was to test novel biodegradable carrier systems, which exhibit greatly improved biocompatibility, for their efficiency in siRNA delivery. By crosslinking of low molecular weight polycations with biodegradable linkers high molecular weight polycations had been created, which can be degraded into smaller fragments in the appropriate cellular microenvironment[128]. This concept was further developed by generating a novel class of hyperbranched polymers, namely pseudodendrimers, which additionally exhibit a better defined chemical structure[135]. These conjugates consist of a low molecular weight polycation, which is functionalized with an excess of biodegradable linker in order to form a pseudodendritic inner core, which can be subsequently modified on its surface with different oligoamines. In an alternative approach, instead of crosslinking, low molecular weight polycations were modified with hydrophobic moieties. These low molecular weight structures had to be characterized as siRNA formulations in their biophysical and biological properties. Conceptually they would possess increased stability against dissociation and enhanced endosomolytic properties, thereby overcoming major bottlenecks for efficient siRNA delivery.

The third aim of this thesis was to evaluate bioresponsive conjugates, which act more dynamically in response to their cellular microenvironment than their static counterparts. As poor endosomal escape is a major barrier of siRNA delivery, polymeric carriers should be tested which are equipped with membrane active peptides, such as melittin, to escape endosomal entrapment. To overcome undesired cytotoxicity in the extracellular compartment due to the general membrane destabilizing properties, the lytic activity of melittin was reversibly masked with a pH responsive protecting group resulting in a triggered lytic activity of melittin only upon acidifaction in the endosome. Additionally, due to the fact that other physiological biomolecules can disrupt siRNA complexes, which results in vector disassembly before reaching the target site, siRNA conjugates should be evaluated that are covalently attached to the carrier system by bioreducible disulfide linkers ensuring release of siRNA in the cytoplasm.

All the mentioned polymer classes and conjugates had to be evaluated in their biophysical (siRNA binding, colloidal stability, release of siRNA) and biological (siRNA delivery efficiency, cytotoxicity, hemolytic activity) properties, which is supposed to reveal promising candidates featuring high efficiency associated with low toxicity for further in vivo studies in tumor bearing mice. Lead candidates had to be analyzed in detail to clarify the effect of charge density, chemical structure, hydrophobicity or lytic activity on their biological properties. In particular, the effect of individual surface modifications on reporter gene silencing efficiency and cytotoxicity had to be investigated and correlated in order to elucidate comprehensive structure-activity relationships. Further studies of the stability and lytic activity of siRNA conjugates should offer some prediction on their fate in extracellular or endosomal environments.

2 MATERIALS AND METHODS

2.1 Chemicals, polymers and other reagents

Oligoethylenimine with an average molecular weight of 800Da (OEI 800), branched polyethylenimine with an average molecular weight of 25kDa (PEI 25), poly-L-lysine-HBr (degree of polymerization = 153) with an average molecular weight of 32kDa as hydrobromide (PLL), succinic anhydride (Suc), 2,3-dimethylmaleic anhydride (DMMAn), N-succinimidyl 3-(2-pyridyldithio)-propionate (SPDP) and 1,4-dithiothreitol (DTT) were obtained from Sigma Aldrich (Steinheim, Germany). Succinimidyl propionate monomethoxy polyethylene glycol with a molecular weight of 5kDa (mPEG5k-SPA) was purchased from Fluka (Buchs, Switzerland). Linear polyethylenimine with an average molecular weight of 22kDa (PEI 22) was synthesized by acid catalysed deprotection of poly(2-ethyl-2-oxazoline) (50kDa) as described in Brissault et al[271] and is also commercially available from Polyplus Transfection (Strasbourg, France).

Cysteine-modified melittin (Mel) was obtained from IRIS Biotech GmbH (Marktredwitz, Germany). Melittin had the sequence CIGAVLKVLTTGLPALISWIKRKRQQ (all-(D) configuration), the C-terminal amino acid was introduced as carboxylic acid, the N-terminal amino acid as amine. All-(D) stereochemistry was used because it is non-immunogenic while being as lytic as the natural all-(L) melittin peptide[272-273].

Transferrin (Tf) was obtained from Biotest (Dreieich, Germany). Deuterium oxide (D_2O), RNAse-free water, absolute ethanol, dimethyl sulfoxide (DMSO), methylthiazolyldiphenyl-tetrazolium bromide (MTT), ethidium bromide (EtBr) and all other chemicals were obtained from Sigma-Aldrich (Steinheim, Germany).

As buffer and solvent were used HBG (HEPES-buffered glucose solution: 20mM HEPES, 5% glucose (w/w), pH 7) or HBS (HEPES-buffered saline: 20mM HEPES, 150mM NaCl, pH 7). Cell culture media, antibiotics and fetal calf serum (FCS) were purchased from Invitrogen (Karlsruhe, Germany).

Ready to use siRNA duplexes were purchased from Dharmacon (Lafayette, CO, USA) and Eurofins MWG Operon (Ebersberg, Germany):

LucsiRNA: GL3 luciferase duplex
5'-CUUACGCUGAGUACUUCGA-3' (sense)
5'-thiol-CUUACGCUGAGUACUUCGA-3' (sense)

2 MATERIALS AND METHODS

RANsiRNA:	RAN specific therapeutic duplex 5'-AGAAGAAUCUUCAGUACUA-3' (sense)
siCONTROL:	non-specific control duplex IX with similar GC content as LucsiRNA: 5'-AUUGUAUGCGAUCGCAGAC-3' non-targeting control duplex siCONTROL#3: 5'-AUGUAUUGGCCUGUAUUAG-3' (sense) 5'-thiol-AUGUAUUGGCCUGUAUUAG-3' (sense)

Following synthetic carrier systems have been used amongst others for siRNA delivery studies:

PEI-EA (13%-26%-52%) **PEI-Prop (13%-26%-52%)** **PEI-Ac (10%-20%)** **PEI-Suc (10%-20%)**	were synthesized as described in Zintchenko et al[160]
OEI-HD-I	was obtained from Abbott Laboratories (Chicago, IL, USA) with an average molecular weight of 25 - 30kDa and was synthesized as described in Kloeckner et al[120] and Tarcha et al[274]
OEI-HD-1-Tf	was synthesized as described in Tietze et al[40]
OEI-ED (E-Sp-S-O) **OEI-BD (E-Sp-S-O)** **OEI-HD (E-Sp-S-O)**	were synthesized as described in Russ et al[135]
OEI-EA (5-10) **OEI-BA (5-10)** **OEI-HA (5-10)** **OEI-LA (2.5-5)**	were synthesized as described in Philipp et al[275]

PEG-PEI PEG-PEI-DMMAn-Mel	were synthesized as described in Meyer et al[247]
PEG-PLL PEG-PLL-DMMAn-Mel	were synthesized as described in Meyer et al[247]
PEG-PLL-DMMAn-Mel-siRNA	was synthesized as described in Meyer et al[85]

For biochemical, biophysical and biological studies, lyophilized conjugates were diluted in RNAse free water and adjusted to pH 7 with HCl.

2.2 Additional novel polymer conjugates

2.2.1 Synthesis of succinic anhydride (Suc) modified OEI-HD-1

OEI-HD-1 (2.0µmol, 50mg) was dissolved in 0.5ml 0.5M NaCl. The desired amount of succinic anhydride was dissolved in DMF and added dropwise to the solution under stirring. The modification degree of Suc to OEI-HD-1 was 5%, 10% and 20% (reagent/amine * 100%, feed). After 24h at room temperature the raw product was concentrated and purified by ultrafiltration (Vivaspin 2, Vivascience, molecular weight cut-off 2000 HY) first three times with 3M NaCl to remove unreacted hydrolysed succinate and then five times with water to remove salt. After purification, the aqueous solution was lyophilized.

The degree of modification with Suc was expressed as a number of modifications per PEI or OEI-HD-1 molecule * 100% (i.e. percentage of modified amines per PEI or OEI-HD-1) and analyzed by ^{1}H-NMR spectroscopy from the ratio between the peaks of ethylene protons of PEI (N$\underline{CH_2CH_2}$, δ 2.6 - 3.6ppm) or OEI-HD-1 (N$\underline{CH_2CH_2}$, δ 2.6 - 3.5ppm) and methylene protons of Suc (HOOC$\underline{CH_2CH_2}$CO, δ 2.5ppm).

2.2.2 Synthesis of PEG modified OEI-HD-1

OEI-HD-1 (1.0µmol, 25mg) dissolved in 1250µl HBS was mixed with mPEG5k-SPA (1.5µmol, 7.5mg) dissolved in 150µl DMSO. After 1h at room temperature the reaction mixture was purified by cation-exchange chromatography (MacroPrep High S; HR 10/10, BioRad, Munich, Germany) and fractionated with a salt gradient from 0.6 to 3.0M NaCl in

20mM HEPES (pH 7). Purification was carried out with a flow rate of 0.5ml/min. The fractions containing PEG-OEI-HD-1 were pooled, dialyzed against water (molecular weight cut-off 6000 - 8000) and lyophilized.

The degree of modification with PEG was determined by ^{1}H-NMR spectroscopy calculated from the proton integrated values of PEG (OCH$_2$CH$_2$, δ 3.6ppm) and the OEI-HD-1 backbone (NCH$_2$CH$_2$, δ 2.6 - 3.5ppm) and from the molecular weight values given by suppliers. The polycation content was determined by TNBS assay. The PEG-OEI-HD-1 conjugate had a molar ratio of OEI-HD-1/PEG = 1/1.15.

2.2.3 Synthesis of 3-(2-pyridyldithio)-propionate modified PEG-OEI-HD-1

PEG-OEI-HD-1 (0.6µmol, 15mg) dissolved 750µl HBS was mixed with SPDP (6µmol, 1.87mg) dissolved in 187µl DMSO. SPDP was used as a heterobifunctional crosslinker reacting with primary and secondary amines via the N-hydroxysuccinimidyl group and reacting with sulfhydryls via the pyridylthiol group.

SPDP

After 2h at room temperature the reaction mixture was purified by size exclusion chromatography using an Äkta Basic high-performance liquid chromatography (HPLC) system (Amersham Biosciences, Freiburg, Germany) equipped with a Sephadex G25 superfine HR 10/30 column (Pharmacia Biotech, Uppsala, Sweden) equilibrated in 0.5M NaCl, 20mM HEPES (pH 7). Gel filtration was carried out with a flow rate of 0.5ml/min and the fractions containing PEG-OEI-HD-1-PDP were pooled.

The degree of modification with PDP was determined spectrophotometrically at 343nm by the release pyridine-2-thione after reduction with 5µl DTT (0.5µmol, 77µg) dissolved in water. The polycation content was determined by TNBS assay. The PEG-OEI-HD-1-PDP conjugate had a molar ratio of PEG-OEI-HD-1/PDP = 1/8.5.

2.2.4 Synthesis of DMMAn-Mel modified PEG-OEI-HD-1

Melittin peptide (1.04μmol, 3mg) dissolved in 200μl buffer (100mM HEPES, 125mM NaOH) was mixed with 500μl ethanol containing DMMAn (7.9μmol, 1mg) by rapid vortexing under argon. After 30min at room temperature the reaction mixture was concentrated and purified by ultrafiltration (Vivaspin 2, Vivascience, molecular weight cut-off 2000 HY) to separate excess of free DMMAn from DMMAn-Mel, which would cause undesired acylation of primary amines on PLL during the coupling procedure.

The final solution of acylated melittin was mixed under argon with PEG-OEI-HD-1-PDP at a 1.5 fold molar excess of DMMAn-Mel (based on PDP content) in 1M guanidine hydrochloride (pH 8) in order to prevent aggregation of the negatively charged DMMAn-Mel and the polycation before coupling. The free sulfhydryl groups of cysteine at the N-terminus of DMMAn-Mel react with PDP to the desired PEG-OEI-HD-1-DMMAn-Mel conjugates.

After 3h at room temperature the release of pyridine-2-thione from residual PDP linkers was measured at 343nm to determine the modification degree. Subsequent purification was carried out by size exclusion chromatography using an Äkta Basic HPLC system (Amersham Biosciences, Freiburg, Germany) equipped with a Superdex 75 HR 10/30 column (Pharmacia Biotech, Uppsala, Sweden) equilibrated in 0.5M NaCl, 20mM HEPES (pH 8) elution buffer to avoid acidic cleavage of DMMAn. Gel filtration was carried out with a flow rate of 0.5ml/min and the fractions containing PEG-OEI-HD-1-DMMAn-Mel were pooled and snap frozen in liquid nitrogen.

The polycation content was determined by TNBS assay. The PEG-OEI-HD-1-DMMAn-Mel conjugate had a molar ratio of PEG-OEI-HD-1/DMMAn-Mel = 1/8.

2.3 Biophysical characterization

2.3.1 siRNA binding ability

The siRNA binding ability of polymers was evaluated using an ethidium bromide exclusion assay. Intercalation of EtBr into plain siRNA results in strongly increased fluorescence of EtBr (λ_{ex}=510nm, λ_{em}=590nm). The ability of polymers to bind siRNA displaces intercalated EtBr, which significantly lowers the fluorescence intensity. Hence, the siRNA binding ability of polymers is expressed as relative fluorescence intensities to plain siRNA with intercalated EtBr, which is set to 100% fluorescence.

Aliquots of the respective polymer were added stepwise to a siRNA solution (20μg/ml) in HBG containing EtBr (0.4μg/ml) and the decrease of fluorescence was measured after each step using a Cary Eclipse fluorescence spectrophotometer (Varian Deutschland GmbH, Darmstadt, Germany).

2.3.2 Polyplex formation

In all studies the composition of polyplexes was characterized by the w/w (weight/weight) ratio of the polymer to the nucleic acid in the mixture. Different concentrations of the polymers and nucleic acid were diluted at various w/w ratios in separate tubes in HBG. Polyplexes were prepared by adding the polymer solution to the solution of the nucleic acid and incubated for 30 minutes at room temperature to form stable complexes.

2.3.3 Agarose gel retardation

Polyplexes were prepared as indicated in the corresponding experimental settings containing 0.5µg siRNA in 20µl HBG. Then complexes were mixed with loading buffer (6ml glycerine, 1.2 ml 0.5M EDTA, 2.8ml H_2O, 0.02g xylenecyanole) and placed into a 2.5% agarose gel in 40g TBE buffer (trizma base 10.8g, boric acid 5.5g, disodium EDTA 0.75g ad 1l water) containing EtBr or GelRed. Electrophoresis was performed at 80V for 40 minutes and evaluated under UV-light.

2.3.4 Polyplex stability against sodium chloride

The stability of polyplexes was studied by dynamic light scattering using a Malvern Zetasizer 3000HS (Malvern Instruments, Worcestershire, UK). A drop of scattering intensity was attributed to the dissociation of polyplexes against increasing amounts of sodium chloride. The NaCl concentration at which the dissociation occurred was related to the polyplex stability.

Polyplexes were prepared with DNA or siRNA as indicated in the corresponding experimental settings at a final nucleic acid concentration of 20µg/ml in HBG. Stepwise addition of 3M NaCl solution to the polyplexes resulted in decrease of scattering intensity as a function of increasing NaCl concentration.

2.3.5 Particle size and zeta-potential measurement

Particle size and zeta-potential of polyplexes were determined using a Zetasizer Nano ZS (Malvern Instruments, Herrenberg, Germany). Polyplexes were prepared with siRNA as indicated in the corresponding experimental settings in HBG. For measurement of zeta-potential polyplexes were diluted with 1mM NaCl to give a final volume of 1ml and a siRNA concentration of 10µg/ml. Dispersion Technology Software 5.0 (Malvern Instruments, Herrenberg, Germany) was used for data acquisition and analysis.

2.3.6 Transmission electron microscopy

For transmission electron microscopy (TEM) investigations 5µl of siRNA conjugates (formulations containing 0.5µg siRNA became diluted 1:100 in water) were put on 3.05mm diameter copper grids with a mesh size of 200µm covered by a 20nm thick lacey carbon film. After air drying of the samples, TEM investigations were performed on a Jeol 2011 microscope equipped with a tungsten filament source and operated with an acceleration voltage of 200kV. Bright field images were recorded on a bottom-mounted CCD camera with a resolution of 1024 x 1024 pixels using a typical exposure time of 1000ms.

To ensure representative results, at least five different areas of 200µm x 200µm were inspected at high magnification on every TEM grid. The contrast is mainly given by the differences in thickness between the actual particle and the surrounding 20nm thick lacey carbon film, as no additional contrast staining was applied.

2.4 Biological characterization

2.4.1 Cell culture

All cultured cells were grown at 37°C in 5% CO_2 humidified atmosphere. Wildtype murine neuroblastoma cells Neuro2A (ATCC CCL-131), Neuro2A/Luc cells stably transfected with the GL3 luciferase gene and Neuro2A/eGFPLuc cells stably transfected with the eGFPLuc gene were cultured in Dulbecco's modified Eagle's medium (DMEM, 1g/l glucose). Wildtype human hepatocellular carcinoma cells HUH7 (JCRB 0403, Tokyo, Japan) and HUH7/eGFPLuc cells stably transfected with the eGFPLuc gene were cultured in DMEM/Ham's F-12 medium. Human lung carcinoma cells H1299/Luc stably transfected with the GL3 luciferase gene (kindly provided by Abbott Laboratories, Chicago, IL, USA) were cultured in RPMI 1640 medium (4.5g/l glucose). All media were supplemented with 10% fetal calf serum (FCS), 2mM stable glutamine, 100U/ml. penicillin and 100µg/ml streptomycin.

2.4.2 Luciferase reporter gene silencing

For screening experiments cells were seeded in 96-well plates (TPP, Trasadingen, Switzerland) using 5000 cells per well and cultured for 24h. Polymer/siRNA complexes containing either LucsiRNA targeting the firefly luciferase or siCONTROL as non-targeting control siRNA (to clearly distinguish between specific gene silencing and unspecific toxic side effects due to the carrier system) were prepared as indicated in the corresponding experimental settings at different w/w ratios in HBG. Prior to siRNA delivery, medium was replaced with 80µl fresh serum containing (10% FCS) growth medium. Then 20µl of polyplex solution was added to each well and cells were incubated at 37°C without further medium change. For transferrin competition experiments, free Tf (iron containing form) in a final concentration of 1µg/µl was added to the cells prior to transfection and medium change was performed 1h following siRNA delivery. At 48h following siRNA delivery the medium was removed and cells were lysed with 50µl of 1:10 diluted cell culture lysis buffer (Promega, Mannheim, Germany).

Luciferase activity was measured using a Lumat LB 9507 Tube Luminometer (Berthold, Bad Wildbad, Germany). Luciferase light units were recorded from a 25µl aliquot of the cell lysate with 10s integration time after automatic injection of 100µl freshly prepared luciferin using the Luciferase Assay System (Promega, Mannheim, Germany). The relative light units (RLU) were expressed as percentage related to untreated control cells.

2.4.3 Metabolic activity of cells after polymer treatment

Metabolic activity of cells after treatment with polymer/siRNA complexes or free polycations was determined using a MTT assay. MTT (3-(4,5-dimethylthiazol-2-yl)-2,5-diphenyltetrazoliumbromide, Sigma-Aldrich, Munich, Germany) was dissolved in phosphate buffered saline (PBS) at 5mg/ml and 10µl aliquots were added to each well reaching a final concentration of 0.5mg MTT/ml. After incubation for 2h at 37°C, the medium was removed and cells were frozen for 1h at -80°C. The purple formazan product was dissolved in 100µl/well dimethyl sulfoxide (DMSO) and quantified by a microplate reader Spectrafluor Plus (Tecan Austria GmbH, Grödig, Austria) at 590nm with background correction at 630nm. Cell viability was expressed as relative metabolic activity related to untreated control cells.

2.4.4 Hemolytic activity of polymers

Murine erythrocytes were isolated from freshly collected citrate buffered blood and washed with PBS by four centrifugation cycles at 2000rpm for 10min at 4°C. The erythrocyte pellet was resuspended in HBG or HBG containing FCS (3% or 10%) at a concentration of 0.1 -

4% (V/V) (~ approximately 10^7 - 10^8 erythrocytes per ml). Polymers were serially diluted in 75µl HBG or HBG containing FCS (3% or 10%) and mixed with 75µl erythrocyte suspension in a V-bottom 96-well plate (NUNC, Roskilde, Denmark). After incubation at 37°C under constant shaking for 10 - 45min, erythrocytes were removed by centrifugation (2000rpm for 10min at 4°C) and 80µl of the supernatant was analyzed for hemoglobin release at 405nm using a microplate reader Spectrafluor Plus (Tecan Austria GmbH, Grödig, Austria).

HBG or HBG containing FCS (3% or 10%) and 1% Triton X-100 were used as negative and positive controls, respectively. Haemolysis was defined as percentage ($OD_{polymer}$ - OD_{buffer})*100 / ($OD_{Triton\ X-100}$ - OD_{buffer}).

2.4.5 Reverse Transcriptase quantitative real-time PCR (RT-qPCR)

For gene expression studies in vitro cells were seeded in 24-well plates (TPP, Trasadingen, Switzerland) using 40000 cells per well and cultured for 24h. Polymer/siRNA complexes were prepared as indicated in the corresponding experimental settings at different w/w ratios in HBG. Prior to siRNA delivery, medium was replaced with 320µl fresh serum containing (10% FCS) growth medium. Then 80µl of polyplex solution was added to each well and cells were incubated at 37°C without further medium change. At 48h following siRNA delivery the medium was removed and cells were lysed with 400µl lysis buffer (Roche Diagnostics, Mannheim, Germany) and homogenized using a syringe and needle. High molecular weight DNA is sheared by passing the lysate through a 20-gauge needle attached to a sterile plastic syringe.

2.4.5.1 RNA isolation and cDNA synthesis

RNA isolation was performed using High Pure RNA Tissue Kit (Roche Diagnostics, Mannheim, Germany) according to the manufacturer's protocol and RNA concentration was determined at a wavelength of 260/280nm using a Biophotometer (Eppendorf, Hamburg, Germany).

First-strand cDNA synthesis was performed with 100ng RNA using Transcriptor High Fidelity cDNA Synthesis Kit (Roche Diagnostics, Mannheim, Germany) according to manufacturer's protocol. Reverse transcription was carried out using random hexamer priming.

2.4.5.2 Quantitative real-time PCR

Dual-colour multiplex real-time analysis was performed on a LightCycler 480 system (Roche Diagnostics, Mannheim, Germany). Primer and probes were designed with the Universal ProbeLibrary (UPL) Assay Design Center using the web-based ProbeFinder software v.2.45 accessible at www.universalprobelibrary.com.

GL3 luciferase:	UPL Probe#29
	Forward primer: 5'-TGAGTACTTCGAAATGTCCGTTC-3'
	Reverse primer: 5'-GTATTCAGCCCATATCGTTTCAT-3'
Mouse RAN:	UPL Probe#2
	Forward primer: 5'-ACCCGCTCGTCTTCCATAC-3'
	Reverse primer: 5'-ATAATGGCACACTGGGCTTG-3'
Mouse ACTB:	UPL Reference Gene Assay (RGA)
Mouse GAPDH:	UPL Reference Gene Assay (RGA)

Mouse ß-actin (ACTB) and glyceraldehyde-3-phosphate dehydrogenase (GAPDH) were used as housekeeping genes and purchased as RGA's from Roche Diagnostics (Mannheim, Germany). Primers for GL3 luciferase and mouse RAN were purchased from Eurofins MWG Operon (Ebersberg, Germany).

Real-time PCR was performed according to the LightCycler 480 system protocol using the following parameters as shown in Table 2.

PCR protocol	Target Temp.	Time
Denaturation	95°C	10min
Amplification (45 cycles)	95°C	10s
	60°C	30s
	72°C	1s
Cooling	40°C	30s

Table 2: PCR parameters for quantitative real-time PCR using the LightCylcer 480 system.

Advanced relative quantification based on the second derivative maximum method was used for data acquisition and analysis performed by the LightCycler 480 quantification software v.3 (Roche Diagnostics, Mannheim, Germany).

2.5 Statistics

Values are presented as mean ± standard deviation and statistical significance of differences was evaluated by one-way analysis of variance (ANOVA). P-values smaller than 0.05 were considered to be significant.

3 RESULTS

3.1 Modified PEIs with reduced toxicity as efficient siRNA carriers

3.1.1 Design of PEI conjugates with reduced charge density

Polymeric carriers such as polyethylenimine (25kDa branched = PEI 25), which proved their efficiency in DNA delivery in vitro and in vivo[100,103] were found to be far less effective in siRNA mediated gene silencing[276-277] and to be rather toxic, when applied in higher concentrations[112-114]. As toxicity is mainly associated with strong positive charges of the polycation, which cause strong interactions with cell surfaces and lastly membrane damage, modifications in order to reduce the positive charges of the polymeric backbone were realized. A number of non-toxic derivates of PEI 25 were generated[160] via modification of amines by ethyl-acrylate, acetylation or introduction of negatively charged propionic acid or succinic acid groups to the polymer structure as shown in Figure 4.

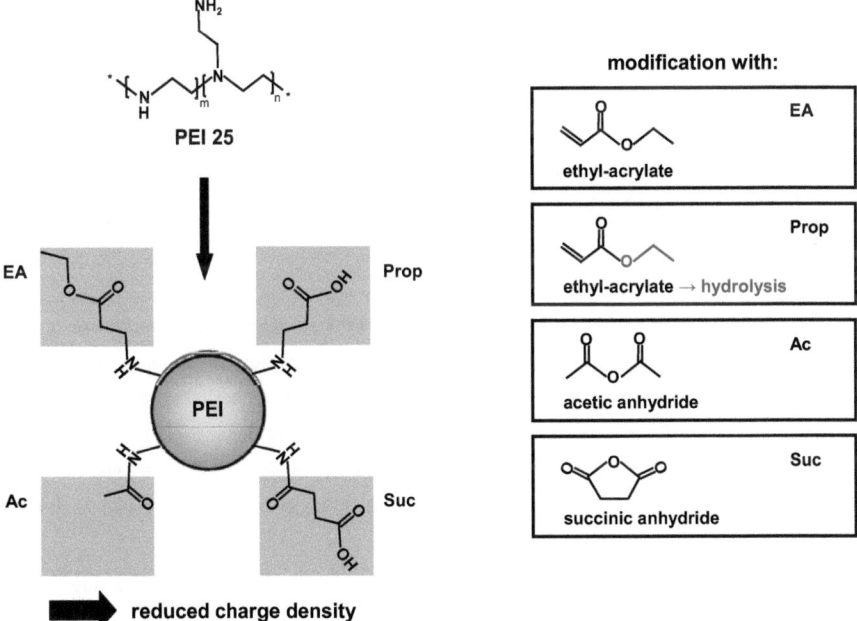

Figure 4: Strategies for modification of PEI 25 resulting in reduced toxicity of the polymers.

3 RESULTS

In the EA and Prop series, primary amino groups were transformed into secondary amino groups by Michael addition, while in the Ac and Suc series, primary amino groups were modified by acetylation into amide groups. Additionally, in the Prop and Suc series negative charges were introduced into the polymer backbone by incorporation of carboxylic groups in the structure.

The structural composition of the resulting products was analyzed using ^{1}H-NMR spectroscopy. The calculated modification degrees (expressed as a percentage of modified amines per PEI 25 molecule) depending on the ratios in the feed are shown in Table 3.

polymer	reagent/amine * 100% (feed)	reagent/amine * 100% (product)
PEI-EA-13%	13%	11.5%
PEI-EA-26%	26%	17.6%
PEI-EA-52%	52%	31%
PEI-Prop-13%	n.a.	11.5%
PEI-Prop-26%	n.a.	17.6%
PEI-Prop-52%	n.a.	31%
PEI-Ac-10%	10%	12.3%
PEI-Ac-20%	20%	21.8%
PEI-Suc-10%	10%	8.9%
PEI-Suc-20%	20%	19.4%

Table 3: Composition of modified PEIs determined by ^{1}H-NMR. The nomenclature of the polymers is derived from the reagent or functional group by which PEI was modified (EA, Prop, Ac, Suc) followed by the modification degree of the amines. Conjugate synthesis was performed by Dr. Arkadi Zintchenko (LMU).

The modification of PEI 25 with ethyl-acrylate via Michael addition was performed at 40°C and a relatively short incubation time of 4h, as higher temperature or longer incubation times are known to cause aminolysis of ester bonds resulting in crosslinking of the polymer[274]. According to the FTIR spectra no amide bond formation and, consequently, no crosslinking was observed for all polymer samples of the PEI-EA series. No double bond peak was found in the ^{1}H-NMR spectra indicating the complete absence of unreacted acrylate in the final products.

The propionic acid modified PEI 25 was generated by acid hydrolysis of the PEI-EA polymers. Complete hydrolysis was verified by disappearance of the methylene protons of the ester bonds in the ^{1}H-NMR spectra and the characteristic FTIR-bands (1730cm^{-1}). The absence of amide peaks in the FTIR spectra (1652 cm^{-1}) reveals no significant crosslinking during hydrolysis. The degree of modification with carboxylic groups was assumed to be the same as in the non-hydrolyzed precursor polymers.

The modification of PEI 25 with acetic anhydride and succinic anhydride was carried out in the presence of salt to avoid precipitation of the polymer. The purification of polymers via dialysis was performed against salt buffer to ensure complete exchange of acetic or succinic acid, against chloride as counterion.

3.1.2 siRNA binding and complexation ability

The capability of polymers to condense siRNA, in order to form complexes suitable for cell entry, was studied using an EtBr exclusion assay. The reduction of relative fluorescence was measured as a function of increasing polymer/siRNA mixing ratios as shown in Figure 5.

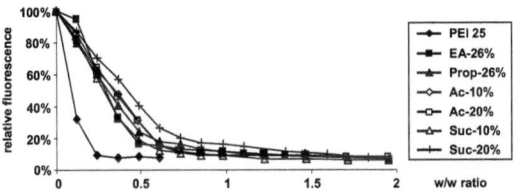

Figure 5: siRNA binding affinity of conjugates determined by EtBr exclusion assay. The numbers on the x-axis represent polymer/siRNA mixing ratios (w/w). Data were generated by Dr. Arkadi Zintchenko (LMU).

Investigating the influence of surface modification on siRNA binding, all polymers were found to be effective in binding of siRNA at low ionic strength buffer HBG. Regarding PEI 25 however, siRNA binding required lower w/w ratios caused by the fact that for PEI 25 the w/w ratio was calculated from the base form in this case. All other modified PEIs were weighed in neutralized form resulting in chloride counter ions which markedly contribute to mass per charge ratio. Nevertheless, at a w/w ratio of 2/1 even polymers with a high degree of modification show a reduction of fluorescence less than 10%.

Agarose gel shift assay (in salt containing TBE buffer), however, illustrates slight differences in binding stability between polymers with different modification degree as shown in Figure 6.

3 RESULTS

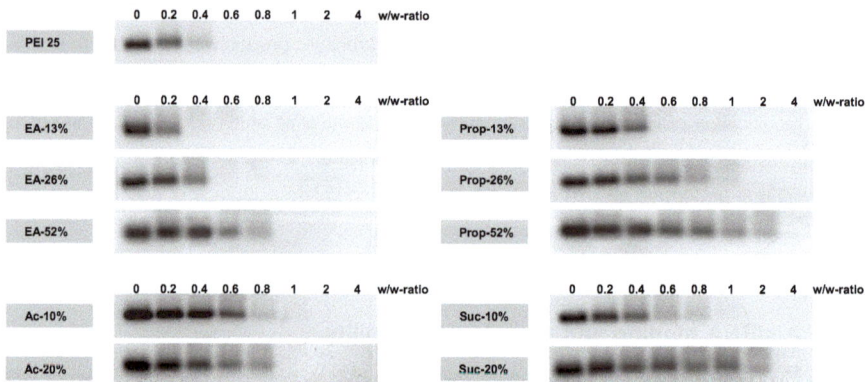

Figure 6: Gel shift assay for polyplexes of siRNA with different modified PEIs. The numbers represent polymer/siRNA mixing ratios (w/w).

Regarding the influence of the modification degree on siRNA binding, it was found that surface modifications with highest modification degree showed lowest siRNA complexation stability, whereas lower modification degrees showed no significant differences in complexation of siRNA in a range of about 0.5 - 0.7 (w/w) compared to each other.

The analysis of polyplex solutions mixed in HBG at w/w ratios of 2 and 4 by photon correlation spectroscopy did not show the formation of nanoparticles in the usual 30 - 300nm range. The scattering intensity from polyplex solutions was close to that of HBG buffer, which indicates very small polyplex sizes and low aggregation behavior of the polymers. An increase of salt concentration to 150mmol NaCl led to continuous aggregation of the initial polyplexes and formation of nanoparticles (around 500nm after 30min) in the case of modified PEIs, whereas unmodified PEI 25 without siRNA did not show any aggregation tendency.

This behavior of PEI/siRNA polyplexes is in accordance with the findings observed by Meyer et al where particle sizes of around 25nm were detected by fluorescent correlation spectroscopy[247].

3 RESULTS

3.1.3 Influence of conjugates on cytotoxicity

To investigate the influence of surface modification on cytotoxicity, the metabolic activity of cells was monitored after treatment with various concentrations of plain polymers as shown in Figure 7.

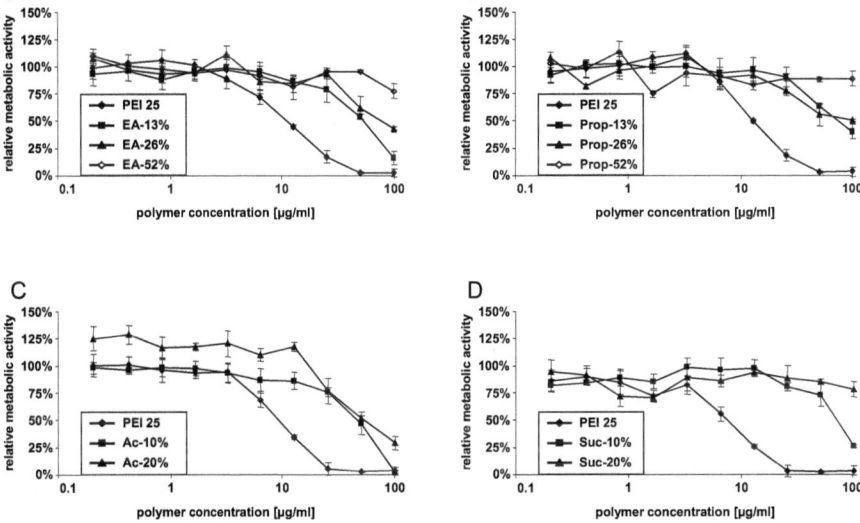

Figure 7: Cell viability of Neuro2A cells monitored by MTT assay as a function of polymer concentration. (A) PEI-EA series, (B) PEI-Prop series, (C) PEI-Ac series, (D) PEI-Suc series.

The toxicity profiles of modified PEIs are shifted, as expected, to higher polymer concentrations in comparison to unmodified PEI 25, demonstrating a decrease in toxicity. Relatively slight modifications (EA and Ac) have shown only moderate improvement on cell viability, whereas the incorporation of negative charges in PEI 25 (Prop and Suc) resulted in far less toxic polymers.

3.1.4 siRNA delivery efficiency: structure-activity relationship

To evaluate the siRNA delivery efficiency of modified polymers, reporter gene silencing studies were carried out using Neuro2A/eGFPLuc, murine neuroblastoma and HUH7/eGFPLuc, human hepatocellular carcinoma cell lines, stably expressing the luciferase reporter gene. Polyplexes were prepared in HBG containing either LucsiRNA targeting the firefly luciferase or siCONTROL as non-targeting control siRNA to clearly distinguish between specific gene silencing and unspecific toxic side effects due to the carrier system. Figure 8 shows the siRNA gene silencing capability of all modified conjugates 48h after initial siRNA delivery on Neuro2A/eGFPLuc cells in serum containing (10% FCS) growth medium.

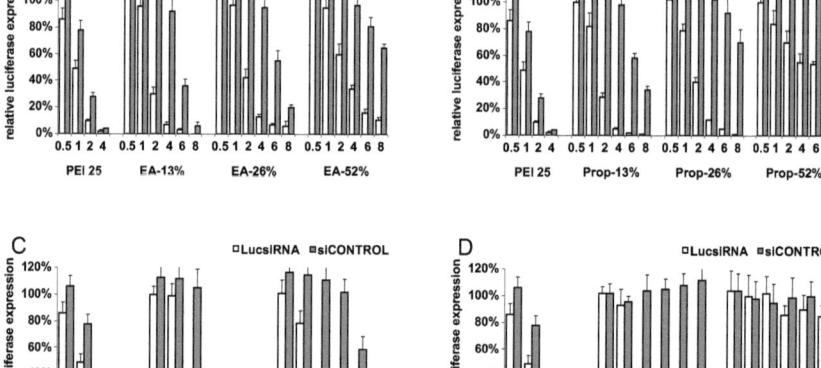

Figure 8: siRNA gene silencing efficiency on Neuro2A/eGFPLuc cells using 500ng siRNA (equal to 380nM). (A) PEI-EA series, (B) PEI-Prop series, (C) PEI-Ac series, (D) PEI-Suc series. White bars indicate complexes containing luciferase siRNA, grey bars indicate complexes containing control siRNA. The numbers on the x-axis represent polymer/siRNA mixing ratios (w/w).

siRNA formulations with unmodified PEI 25 were not able to mediate knockdown of luciferase expression without unspecific toxicity. High toxicity of PEI 25 revealed at w/w ratios higher than 1/1 indicated by the decrease of luciferase expression for formulations with control siRNA.

In contrast, siRNA delivery efficiency of most modified PEIs (except PEI-Prop-52% and PEI-Suc-20%) was greatly enhanced as demonstrated by luciferase knockdown up to 80 - 90%, while luciferase levels for formulations with control siRNA remained unaffected. Interestingly, PEI-EA-52% showed significant knockdown at w/w ratios of 4/1 and 6/1, whereas the hydrolyzed analogue PEI-Prop-52% was inactive. Thus, low amounts of negatively charged carboxylic groups on the polymer backbone were able to improve the knockdown efficiency of the formulation, whereas high amounts of negative charges resulted in non-toxic but also non-effective polymers.

The best knockdown effect was achieved by PEI-Suc-10% demonstrating gene silencing effects of even > 90%, while being non-toxic at all studied mixing ratios. However, further

3 RESULTS

increase of the negative charges led to inactivation of the polymer, as PEI-Suc-20% was completely ineffective for luciferase knockdown at all w/w ratios.

Even at lower amounts of siRNA, a significant knockdown was achieved, if the polymer amount was kept constant as shown in Figure 9.

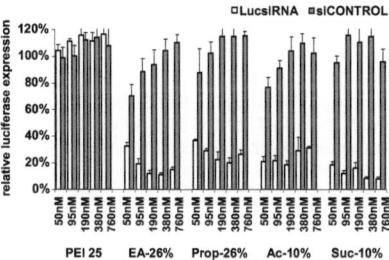

Figure 9: siRNA gene silencing efficiency on Neuro2A/eGFPLuc cells using 62.5ng - 1000ng siRNA (equal to 50nM - 760nM). The concentration of unmodified PEI 25 was kept constant at 2.5µg/ml, the concentration of modified PEIs was kept constant at 20µg/ml. White bars indicate complexes containing luciferase siRNA, grey bars indicate complexes containing control siRNA. The numbers on the x-axis represent the siRNA concentration.

When polymer concentrations were fixed at 20µg/ml, modified PEIs achieved 60 - 80% knockdown of luciferase expression even at low siRNA concentrations of just 50nM. The decrease of siRNA concentration led to a certain increase of toxicity caused by an increasing amount of uncomplexed polymer in the transfection medium. Only PEI-Suc-10% was able to show up to 80% knockdown without any sign of toxicity even at 50nM siRNA.

Unmodified PEI 25 at a concentration of 2.5µg/ml was used as a control which remained completely ineffective at any siRNA concentration. Higher polymer concentrations showed unspecific knockdown effects due to unspecific toxicity of the carrier system.

PEI-Suc-10%, as the most promising candidate, was additionally tested for siRNA delivery on HUH7/eGFPLuc cells as shown in Figure 10.

3 RESULTS

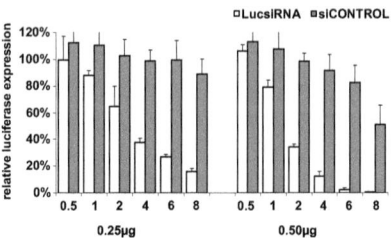

Figure 10: siRNA gene silencing efficiency of PEI-Suc-10% on HUH7/eGFPLuc cells using 250ng or 500ng siRNA (equal to 190nM or 380nM). White bars indicate complexes containing luciferase siRNA, grey bars indicate complexes containing control siRNA. The numbers on the x-axis represent polymer/siRNA mixing ratios (w/w).

Also on this cell line, PEI-Suc-10% showed greatly enhanced siRNA delivery efficiency demonstrated by excellent marker gene knockdown in a dose dependant manner. At lower siRNA concentrations (190nM) more polymer, i.e. higher w/w-ratios of the PEI-Suc-10%/siRNA complex is needed for efficient gene silencing. Moreover, a slightly toxic effect of the polymer could be observed on this cells line at a high w/w ratio of 8/1.

3.1.5 Study on mechanism of the highly effective siRNA carrier PEI-Suc-10%

Modified PEI formulations were superior in efficiency and toxicity compared with unmodified PEI 25. This difference is caused either directly by a different interaction of the polymer with siRNA in the polyplex, or indirectly, by the presence of additional free polymer which is separately internalized into endocytic vesicles[278].

To analyze this issue, siRNA polyplexes were formed at lower (ineffective) w/w ratios with either PEI-Suc-10% or unmodified PEI 25. At 1h after siRNA delivery free PEI-Suc-10% was added separately to the cells. As shown in Figure 11A, the addition of free PEI-Suc-10% strongly increases the siRNA delivery efficiency of PEI-Suc-10%/siRNA complexes.

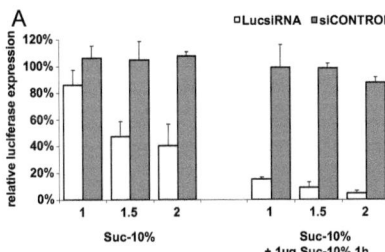 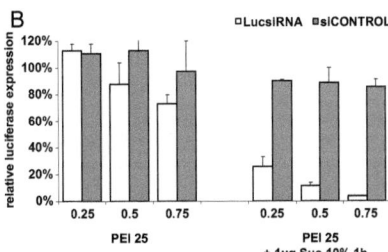

Figure 11: siRNA gene silencing efficiency on Neuro2A/eGFPLuc cells using 500ng siRNA (equal to 380nM). (A) PEI-Suc-10%/siRNA complexes, (B) PEI 25/siRNA complexes. At 1h after siRNA delivery 1µg free PEI-Suc-10% was added separately to the cells where specified. White bars indicate complexes containing luciferase siRNA, grey bars indicate complexes containing control siRNA. The numbers on the x-axis represent polymer/siRNA mixing ratios (w/w).

When using unmodified PEI 25/siRNA complexes similar observations were made as shown in Figure 11B. While the unmodified polyplexes were ineffective, subsequent addition of free PEI-Suc-10% resulted in effective knockdown of luciferase expression.

Thus, a better endosomal escape by the increased amount of the less toxic modified polymer seems to be the main reason for the excellent knockdown effect for modified PEIs. Application of the same high amount of PEI 25 would be cytotoxic (see Figure 8).

3.1.6 Polyplex stability against sodium chloride induced dissociation

The influence of PEI modifications on stability of polyplexes was studied using photon correlation spectroscopy. Electrostatic interactions between polycations and nucleic acids, largely responsible for the stability of polyplexes, can be disturbed by salt resulting in dissociation of polyplexes long before they reach their target site of action. Hence, dynamic light scattering was used to monitor complex disassembly as a function of increasing NaCl concentration, as intensity of scattered light is dramatically decreased when complexes dissociate. The salt concentration required for the dissociation was attributed to the stability of polyplexes as shown in Figure 12.

3 RESULTS

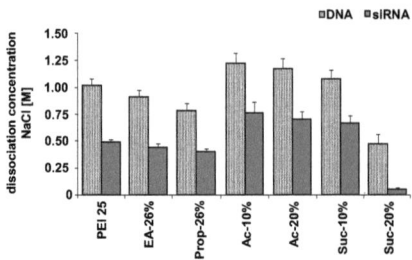

Figure 12: NaCl concentrations required for dissociation of polyplexes. Bright bars indicate complexes containing DNA, dark bars indicate complexes containing siRNA prepared at a w/w ratio of 1/1 and a final nucleic acid concentration of 20µg/ml. Data were generated by Dr. Arkadi Zintchenko (LMU).

Polyplexes with DNA appeared to be more stable than polyplexes with siRNA as dissociation occurred at nearly two times higher salt concentrations. However, no simple correlation between the stability and knockdown efficiency of polyplexes generated with siRNA was found, since all series of PEI modifications showed efficient knockdown of luciferase expression.

According to the stability data, EA and Prop series showed lower stability than unmodified PEI 25 polyplexes, whereas Ac and Suc series showed higher stability of polyplexes. For both, DNA and siRNA polyplexes, the same tendency was observed.

The increased colloidal stability in case of Ac and Suc series might be due to additional stabilization of polyplexes via hydrogen bonding between the amides in the polymer structure and the nucleic acids[279].

As an exception, PEI-Suc-20%, which is relatively stable with DNA, was found to be entirely instable with siRNA at physiological salt concentrations and hence not able to induce efficient siRNA mediated knockdown (see Figure 8).

3.2 Biodegradable OEI conjugates for siRNA delivery

Polyethylenimine is one of the most studied and commonly used polycations for nucleic acid delivery[100]. However, representing a non-degradable high molecular weight polymer, insufficient metabolization and elimination result in undesired short- and long-term toxicity in vivo due to a variety of unspecific interactions with the biological environment and unintentional accumulation in cells and excretion organs, such as liver.

Accordingly, our lab and other investigators have generated novel biodegradable polymer conjugates based on low molecular weight oligoethylenimine (OEI). These conjugates are designed to decompose into low molecular weight fragments which are less toxic and can be more easily eliminated from the organism. Furthermore, crosslinkers may be applied which actively improve intracellular release of the nucleic acid from the carrier system, resulting in enhanced delivery properties of polyplexes.

3.2.1 Polymers based on oligomerized OEIs (OEI-HD-1)

3.2.1.1 Design of OEI-HD-1

One approach to generate biodegradable high molecular weight conjugates is based on low molecular weight oligomers with low toxicity, which are crosslinked to form larger polycationic carriers by conjugation with biodegradable linkers[119-120,123,127-129,132-133]. Thus, a more efficient and potentially degradable oligoethylenimine-based carrier system for siRNA delivery was synthesized by Michael addition of OEI 800 oligomers with hexanediol-diacrylate at a molar ratio of 1/1[128,274]. Reaction was followed by complete N-acylation of all residual ester bonds resulting in beta-aminopropionamide linkages as shown in Figure 13.

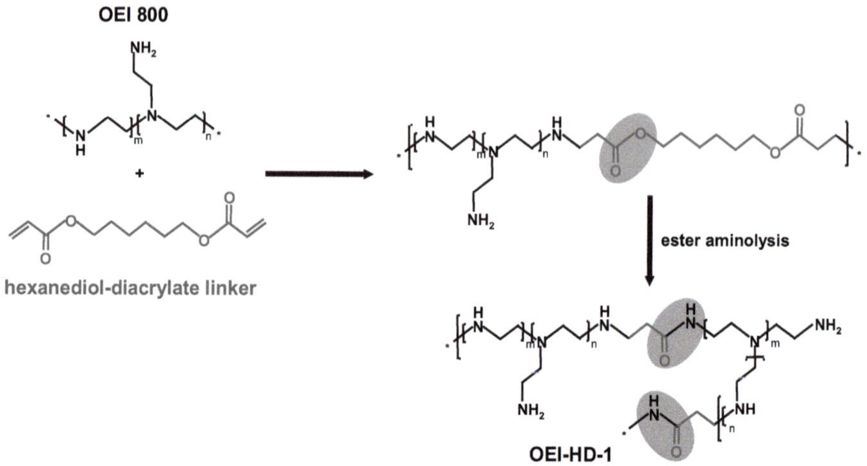

3 RESULTS

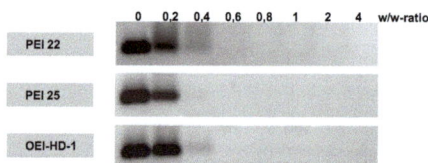

Figure 13: Synthesis of OEI-HD-1. Oligoethylenimine was coupled with hexanediol-diacrylate at a molar ratio of 1/1 in anhydrous DMSO followed by N-acylation of resulting ester bonds. Conjugate synthesis was developed by Julia Klöckner as part of her PhD thesis (LMU, 2005) and optimized by Dr. Peter Tarcha (Abbott).

3.2.1.2 siRNA complexation ability

The ability of OEI-HD-1 to complex siRNA in HBG was confirmed by agarose gel shift assay (in salt containing TBE buffer) as shown in Figure 14.

Figure 14: Gel shift assay for polyplexes of siRNA with PEI 22, PEI 25 and OEI-HD-1. The numbers represent polymer/siRNA mixing ratios (w/w).

OEI-HD-1 is able to completely complex siRNA at a w/w ratio of 0.5/1 or higher in analogous manner to linear PEI 22 and branched PEI 25.

Furthermore, particle sizes of complexes prepared in HBG at w/w ratios of 1/1 and 2/1 were determined by photon correlation spectroscopy and resulted in multimodal distribution (10 - 1000 nm)[40], whereas low scattering intensity implies that predominantly smaller particles are present.

This behavior of OEI-HD-1/siRNA polyplexes is in accordance with the findings observed by Meyer et al where particle sizes of around 25nm were detected by fluorescent correlation spectroscopy[247].

3.2.1.3 siRNA delivery efficiency and toxicity of OEI-HD-1

To evaluate the siRNA delivery efficiency of OEI-HD-1, reporter gene silencing studies were carried out using Neuro2A/eGFPLuc, murine neuroblastoma and HUH7/eGFPLuc, human hepatocellular carcinoma cell lines, stably expressing the luciferase reporter gene. Polyplexes were prepared in HBG containing either LucsiRNA targeting the firefly luciferase or siCONTROL as non-targeting control siRNA to clearly distinguish between specific gene silencing and unspecific toxic side effects due to the carrier system. Figure 15A shows the siRNA gene silencing capability of linear PEI 22, branched PEI 25 and OEI-HD-1 conjugates 48h after initial siRNA delivery on Neuro2A/eGFPLuc cells in serum containing (10% FCS) growth medium.

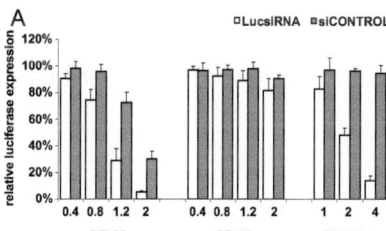
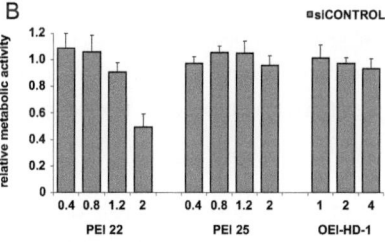

Figure 15: siRNA gene silencing efficiency on Neuro2A/eGFPLuc cells (A) and cell viability of Neuro2A cells monitored by MTT assay (B) using 250ng siRNA (equal to 190nM). White bars indicate complexes containing luciferase siRNA, grey bars indicate complexes containing control siRNA. The numbers on the x-axis represent polymer/siRNA mixing ratios (w/w).

siRNA formulations with linear PEI 22 displayed only minor reduction of luciferase expression, as the knockdown effect is mainly associated with toxic side effects of the carrier system. PEI 25 in contrast, did not mediate any significant reduction in luciferase activity. Only OEI-HD-1 showed a remarkable knockdown effect at a w/w ratio of 4/1 in the absence of unspecific toxicity resulting in reporter gene silencing > 80% compared to untreated control cells.

Figure 15B shows the influence on metabolic activity of Neuro2A cells after polyplex treatment under same conditions, which is consistent with the unspecific reduction of luciferase expression (siCONTROL formulations) on Neuro2A/eGFPLuc cells.

OEI-HD-1 was additionally tested for siRNA delivery on HUH7/eGFPLuc cells as shown in Figure 16.

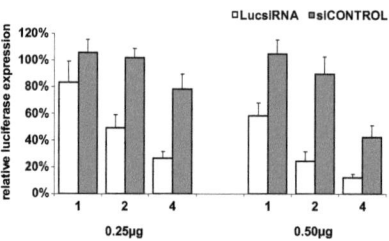

Figure 16: siRNA gene silencing efficiency of OEI-HD-1 on HUH7/eGFPLuc cells using 250ng or 500ng siRNA (equal to 190nM or 380nM). White bars indicate complexes containing luciferase siRNA, grey bars indicate complexes containing control siRNA. The numbers on the x-axis represent polymer/siRNA mixing ratios (w/w).

Also on this cell line, OEI-HD-1 showed pronounced knockdown of luciferase expression without unspecific toxicity. When using 500ng siRNA, the optimal OEI-HD-1/siRNA ratio, regarding maximal efficiency at minimal toxicity, was shifted to a lower w/w ratio of 2/1 while significant toxicity occurred at ratios greater 2/1.

Moreover, in vitro reporter gene silencing of OEI-HD-1/siRNA complexes was also evaluated in 24-well plates (40000 cells per well) using both the luciferase reporter gene assay and RT-qPCR analysis (Figure 17) in order to quantify luciferase knockdown directly on mRNA level.

3 RESULTS

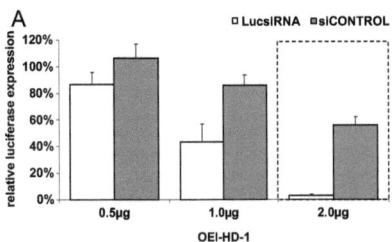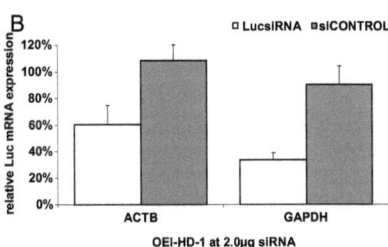

Figure 17: siRNA gene silencing efficiency on Neuro2A/Luc cells (40000 cells per well) following treatment with OEI-HD-1/siRNA complexes at a w/w ratio of 2/1. (A) Luciferase expression measured by luciferase reporter gene assay using 500ng - 2000ng siRNA (equal to 95nM - 380nM). (B) Relative quantification of mRNA levels measured by RT-qPCR analysis using 2000ng siRNA (equal to 380nM). Luc mRNA levels were normalized to expression levels of housekeeping genes ACTB and GAPDH. White bars indicate complexes containing luciferase siRNA, grey bars indicate complexes containing control siRNA.

Polyplexes, prepared at a w/w ratio of 2/1 using 2µg siRNA per 40000 cells showed strongest knockdown of luciferase expression in the luciferase reporter gene assay (Figure 17A). According to the RT-qPCR analysis specific knockdown of luciferase mRNA levels by polyplex treatment was also detected against the housekeepers ACTB and GAPDH (Figure 17B) indicating that knockdown of luciferase is due to the RNA interference mechanism and not due to any unspecific interactions.

In order to evaluate the potential therapeutic effect of siRNA mediated gene silencing in tumor cells, the endogenous RAS related nuclear protein RAN was selected as a therapeutically relevant target[280-281]. As the small GTPase RAN is involved in the regulation of nuclear transport and spindle assembly[282-283], downregulation with RANsiRNA is expected to reduce the survival of tumor cells due to apoptotic cell death. Figure 18A shows the metabolic activity of wild type Neuro2A cells in 24-well plates (40000 cell per well) after treatment with OEI-HD-1 formulations at a w/w ratio of 2/1 containing RAN specific siRNA and control siRNA, respectively.

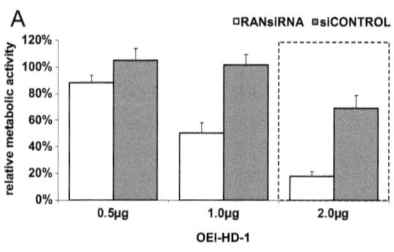 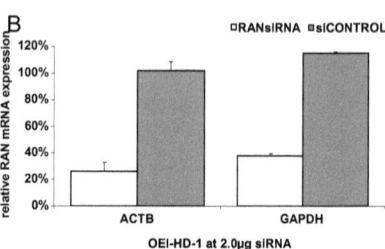

Figure 18: siRNA gene silencing efficiency on Neuro2A cells (40000 cells per well) following treatment with OEI-HD-1/siRNA complexes at a w/w ratio of 2/1. (A) Cell viability measured by MTT assay using 500ng - 2000ng siRNA (equal to 95nM - 380nM). (B) Relative quantification of mRNA levels measured by RT-qPCR analysis using 2000ng siRNA (equal to 380nM). RAN mRNA levels were normalized to expression levels of housekeeping genes ACTB and GAPDH. White bars indicate complexes containing RANsiRNA, grey bars indicate complexes containing control siRNA.

In this approach RANsiRNA treated cells showed strongly reduced cell viability in contrast to the cells treated with control siRNA formulations reflecting the potential therapeutic effect of RANsiRNA mediated knockdown in tumor cells. RT-qPCR analysis also revealed pronounced knockdown of RAN mRNA levels against the housekeepers ACTB and GAPDH within RANsiRNA treated cells, whereby no alterations in the mRNA levels for the control treated and non-treated cells could be detected (Figure 18B).

3.2.1.4 Transferrin receptor targeting of siRNA polyplexes

In order to enhance the specifity of polyplexes towards tumor cells, the serum protein transferrin was incorporated as a targeting ligand into OEI-HD-1 conjugates. Besides cell specific uptake via the Tf-receptor, transferrin additionally acts as a surface shielding agent optimizing siRNA polyplexes for systemic application in vivo[193-194].

Transferrin conjugated OEI-HD-1 was synthesized by Wolfgang Rödl (LMU) as described in Tietze et al[40].

In preliminary studies optimized transferrin containing OEI-HD-1 formulations were figured out consisting of 10 weight percentage of OEI-HD-1-Tf and 90 weight percentage of OEI-HD-1, whereby the amounts were calculated in relation to the weight of unmodified OEI-HD-1[40]. Notably, these targeted OEI-HD-1-Tf (10%)/siRNA complexes displayed lower zeta potentials compared to non-targeted formulations and effective reporter gene silencing in Neuro2A/eGFPLuc cells, thus appearing more suitable for in vivo applications[40].

To demonstrate Tf functionality of the targeted OEI-HD-1-Tf (10%)/siRNA formulations, Tf-receptor expressing Neuro2A/eGFPLuc cells were saturated with free Tf prior to siRNA delivery as shown in Figure 19.

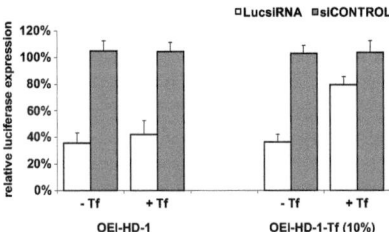

Figure 19: siRNA gene silencing efficiency on Neuro2A/eGFPLuc cells using 500ng siRNA (equal to 380nM). Prior to siRNA delivery, free Tf in a final concentration of 1µg/µl was added to the cells where specified. White bars indicate complexes containing luciferase siRNA, grey bars indicate complexes containing control siRNA. Complexes were prepared at a w/w ratio of 2/1 and 1h after siRNA delivery medium change was performed.

OEI-HD-1/siRNA formulations showed knockdown of luciferase expression independent of the presence of free Tf. In contrast, knockdown efficiency of targeted OEI-HD-1-Tf (10%)/siRNA formulations was reduced to 20% when free Tf was in the medium. These experiments indicate the functionality of Tf ligand as part of the polyplexes making these formulations very interesting for further in vivo investigations performed by Nicole Tietze as part of her PhD thesis (LMU, 2009).

3.2.1.5 Succinylated OEI-HD-1 for improved effective window

Based on the encouraging findings on improved siRNA delivery with succinylated PEI 25, this promising modification strategy was extended to OEI-HD-1 taking advantage of its biodegradability and associated low long term toxicity.

Succinylation of OEI-HD-1 was carried out analogously to PEI 25 as shown in Figure 20.

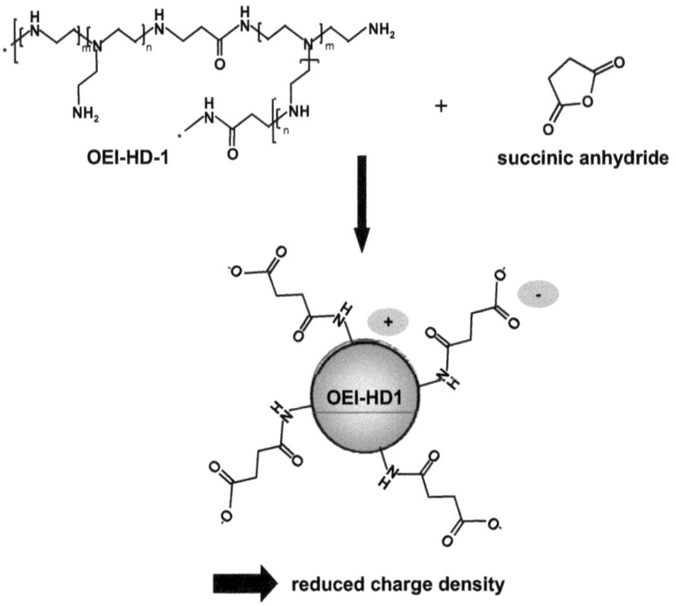

Figure 20: Succinylation of OEI-HD-1 resulting in reduced toxicity of the polymer.

The structural composition of the resulting products was analyzed using ^{1}H-NMR spectroscopy. The calculated modification degrees (expressed as a percentage of modified amines per OEI-HD-1 molecule) depending on the ratios in the feed are shown in Table 4.

polymer	Suc/amine * 100% (feed)	Suc/amine * 100% (product)	Suc/OEI mol/mol (product)
OEI-HD-1-Suc-5%	5%	7.8%	1.4
OEI-HD-1-Suc-10%	10%	12.0%	2.3

Table 4: Composition of succinylated OEI-HD-1 determined by ^{1}H-NMR. The nomenclature of the polymers is derived from the reagent by which OEI-HD-1 was modified (Suc) followed by the modification degree of the amines. Conjugate synthesis was performed by Dr. Arkadi Zintchenko (LMU).

The modification degree was additionally expressed as a number of succinate molecules per OEI 800 unit (Suc/OEI mol/mol). It was previously found that only primary amines are able to react with succinic anhydride in salt containing buffer. Since OEI 800 has around 7 primary amines (of approximately 18), 4 of those are already modified during oligomerisation by hexanediol-diacrylate, the Suc/OEI ratio gets saturated around a value of 3.

3 RESULTS

The ability of modified OEI-HD-1 to complex siRNA in HBG was confirmed by agarose gel shift assay (in salt containing TBE buffer) as shown in Figure 21.

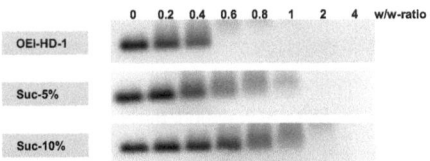

Figure 21: Gel shift assay for polyplexes of siRNA with succinylated OEI-HD-1. The numbers represent polymer/siRNA mixing ratios (w/w).

To evaluate the siRNA delivery efficiency of modified OEI-HD-1, reporter gene silencing studies were carried out using Neuro2A/eGFPLuc, murine neuroblastoma cell line, stably expressing the luciferase reporter gene. Polyplexes were prepared in HBG containing either LucsiRNA targeting the firefly luciferase or siCONTROL as non-targeting control siRNA to clearly distinguish between specific gene silencing and unspecific toxic side effects due to the carrier system. Figure 22 shows the siRNA gene silencing capability of modified OEI-HD-1 conjugates 48h after initial siRNA delivery on Neuro2A/eGFPLuc cells in serum containing (10% FCS) growth medium.

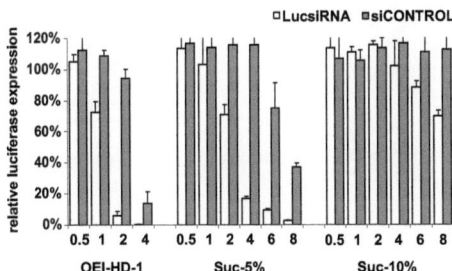

Figure 22: siRNA gene silencing efficiency on Neuro2A/eGFPLuc cells using 500ng siRNA (equal to 380nM). White bars indicate complexes containing luciferase siRNA, grey bars indicate complexes containing control siRNA. The numbers on the x-axis represent polymer/siRNA mixing ratios (w/w).

Succinylation of OEI-HD-1 generally resulted in far less toxic formulations in comparison to unmodified precursor molecules. However the efficient knockdown of luciferase expression was only observed for OEI-HD-1-Suc-5% formulations, whereby the optimal ratio for gene silencing was slightly shifted to higher w/w ratios. Thus, as expected, succinylation of OEI-HD-1 is able to increase the therapeutic window of the formulation.

3.2.2 Pseudodendritic oligoamines

3.2.2.1 Design of OEI core based conjugates

Pseudodendrimers represent a novel class of biodegradable branched polymers, which exhibit a better defined chemical structure compared to commonly synthesized polycations[135]. Furthermore, consisting of a pseudodendritic core, which results from a low molecular weight polycation, e.g. oligoethylenimine, functionalized with an excess of degradable dioldiacrylate linker, they can be further modified on the surface due to free linker ends with different charge bearing compounds, as shown in Figure 23.

3 RESULTS

Figure 23: Concept of pseudodendrimer synthesis. Pseudodendritic core formation, using oligoethylenimine coupled with different dioldiacrylates at a 20-fold molar excess of linker in anhydrous DMSO, was followed by surface modification, using different oligoamines at a 30-fold molar excess of oligoamine to core OEI in anhydrous DMSO. Conjugate synthesis and further characterization was performed by Verena Russ as part of her PhD thesis (LMU, 2008).

Pseudodendrimers with different cores and different surface modifications were synthesized in a two-step procedure. First, pseudodendritic cores were generated by Michael addition of OEI 800 oligomers with different dioldiacrylates (increasing the core hydrophobicity: ED, BD, HD) at a 20-fold molar excess of linker[135]. Synthesis was carried out at 45°C for 24h, while the excess of dioldiacrylate linker prevented crosslinking of the polymer resulting in a branched structure with free linker ends. Second, OEI cores were further modified on the surface with different oligoamines (increasing the number of nitrogens on surface: E, Sp, S, O) at a 30-fold molar excess of oligoamine to core OEI[135]. Synthesis was carried out at 22°C for 24h.

The nomenclature of pseudodendrimers is derived from the core moiety (ED, BD, HD) followed by the modification on the surface (E, Sp, S, O).

3.2.2.2 siRNA delivery efficiency: structure-activity relationship

To evaluate the siRNA delivery efficiency of pseudodendrimers, reporter gene silencing studies were carried out using Neuro2A/eGFPLuc, murine neuroblastoma and HUH7/eGFPLuc, human hepatocellular carcinoma cell lines, stably expressing the luciferase reporter gene. Polyplexes were prepared in HBG containing either LucsiRNA targeting the firefly luciferase or siCONTROL as non-targeting control siRNA to clearly distinguish between specific gene silencing and unspecific toxic side effects due to the carrier system. Figure 24 shows the siRNA gene silencing capability of all pseudodendrimers 48h after initial siRNA delivery on Neuro2A/eGFPLuc cells in serum containing (10% FCS) growth medium.

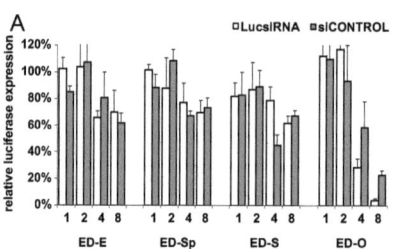
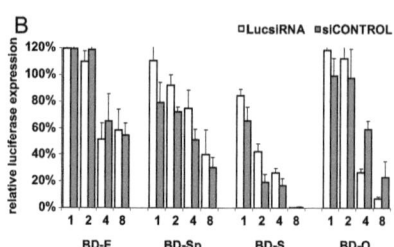
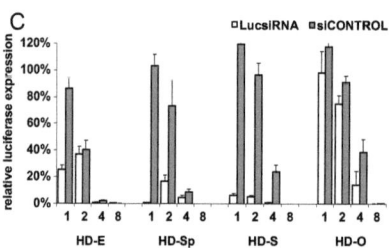

Figure 24: siRNA gene silencing efficiency on Neuro2A/eGFPLuc cells using 500ng siRNA (equal to 380nM). (A) OEI-ED series, (B) OEI-BD series, (C) OEI-HD series. White bars indicate complexes containing luciferase siRNA, grey bars indicate complexes containing control siRNA. The numbers on the x-axis represent polymer/siRNA mixing ratios (w/w).

Screening studies revealed OEI-HD-Sp and OEI-HD-S as the only effective conjugates for siRNA delivery.

Regarding the effects of the different pseudodendritic core characteristics on siRNA delivery, increased toxicity was detected for siRNA complexes formed with OEI-HD core conjugates over the OEI-BD and OEI-ED core conjugates which is accompanied with increasing core hydrophobicity (ED < BD < HD).

Referred to the influences of the various surface modifications on siRNA delivery, it was found that only Sp and S modifications within the OEI-HD core conjugates were able to mediate efficient knockdown of luciferase expression without unspecific toxicity.

OEI-HD-Sp and OEI-HD-S were additionally tested for siRNA delivery on HUH7/eGFPLuc cells as shown in Figure 25.

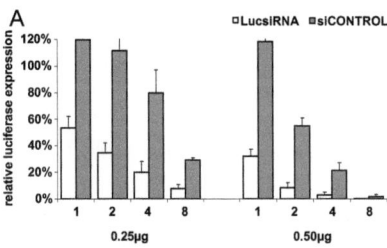 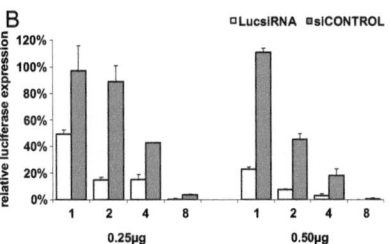

Figure 25: siRNA gene silencing efficiency on HUH7/eGFPLuc cells using 250ng or 500ng siRNA (equal to 190nM or 380nM). (A) OEI-HD-Sp/siRNA complexes, (B) OEI-HD-S/siRNA complexes. White bars indicate complexes containing luciferase siRNA, grey bars indicate complexes containing control siRNA. The numbers on the x-axis represent polymer/siRNA mixing ratios (w/w).

Both conjugates showed again efficient knockdown of luciferase expression in the absence of unspecific toxicity.

Apparently, efficient reporter gene silencing seems to be dependent on an optimized balance of core characteristics and the surface amines. The number of nitrogens upon the pseudodendritic OEI-HD core seem to be optimized in case of Sp (3 N per unit) and S (4 N per unit) exhibiting the best gene silencing effect.

3.2.3 Hydrophobically modified OEIs

3.2.3.1 Design of modified OEI conjugates

Polymeric carrier systems are often based on larger macromolecules, contingently resulting in accumulation of toxicity and narrow therapeutic windows, as low molecular weight polycations are known to be quite ineffective, but also non-toxic. Thus, another approach was evaluated: instead of covalent crosslinking into high molecular structures, the low molecular weight oligoamine OEI was hydrophobically modified in order to ensure short half-life times in the organism, which is favourable for elimination[275]. Moreover, hydrophobic interactions are supposed to stabilize polyplexes during storage and administration and especially to enhance cell membrane interactions promoting quick release out of the endosomes. Thus, a biodegradable oligoethylenimine-based carrier system for siRNA delivery was synthesized by Michael addition of OEI 800 oligomers with different alkyl-acrylates as shown in Figure 26.

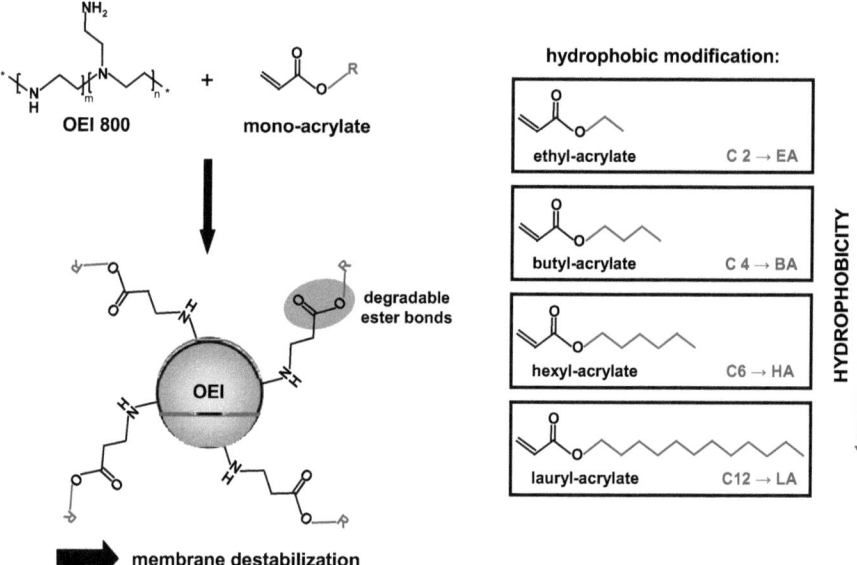

Figure 26: Structure of hydrophobically modified OEIs mediating steric stabilization of polyplexes and causing cell membrane destabilization.

The structural composition of the resulting products was analyzed using ^{1}H-NMR spectroscopy. The calculated modification degrees (expressed as a number of modifications per OEI 800 molecule) depending on the ratios in the feed are shown in Table 5.

3 RESULTS

polymer	reagent/OEI mol/mol (feed)	reagent/OEI mol/mol (product)
OEI-EA-5	5	4.75
OEI-EA-10	10	9.03
OEI-BA-5	5	4.81
OEI-BA-10	10	9.44
OEI-HA-5	5	5.34
OEI-HA-10	10	10.5
OEI-LA-2.5	2.5	2.37
OEI-LA-5	5	5.4

Table 5: Composition of modified OEIs determined by ¹H-NMR. The nomenclature of the polymers is derived from the reagent by which OEI was modified (EA, BA, HA, LA) followed by the modification degree of the amines. Conjugate synthesis was performed by Dr. Arkadi Zintchenko (LMU).

The modification of OEI 800 with alkyl-acrylates via Michael addition was performed at 40°C and a relatively short incubation time of 4h, as higher temperature or longer incubation time are known to cause aminolysis of ester bonds resulting in crosslinking of the polymer[274]. According to the FTIR spectra no amide bond formation and, consequently, no crosslinking was observed for all polymer samples. No double bond peak was found in the ¹H-NMR spectra indicating the complete absence of unreacted acrylate in the final products.

Further advantage of such structures (in comparison to modification with alkyl-acrylamides) is relatively rapid enzymatic degradation of ester bonds in the body and renal clearance of metabolites. Hence, degradation studies for hydrophobically modified conjugates were carried out at physiological pH of 7 at 37°C. The extent of degradation was determined by ¹H-NMR spectroscopy as shown in Figure 27.

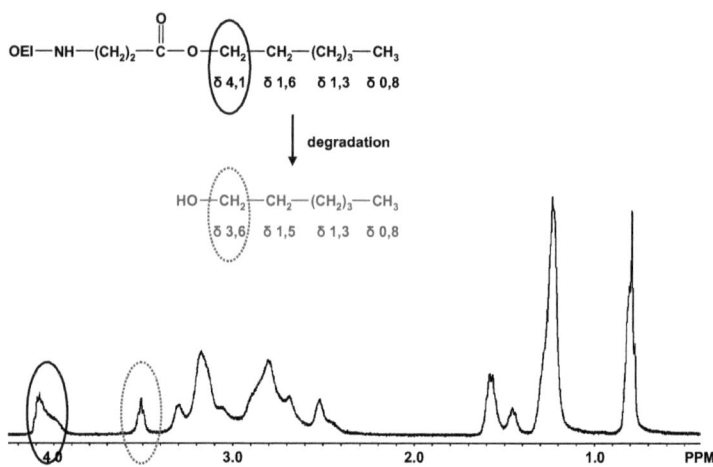

Figure 27: ^{1}H-NMR spectrum of the degradation product of OEI-HA-10 after 3 days incubation at physiological pH of 7 at 37°C.

Even in pH neutral buffer (without enzymes) after an incubation period of 3 days, OEI-HA-10 revealed up to 30% degradation of ester bonds indicating low long-term toxicity of the conjugate.

3.2.3.2 siRNA complexation ability

The influence of surface modification on siRNA binding stability was investigated by agarose gel shift assay (in salt containing TBE buffer) as shown in Figure 28.

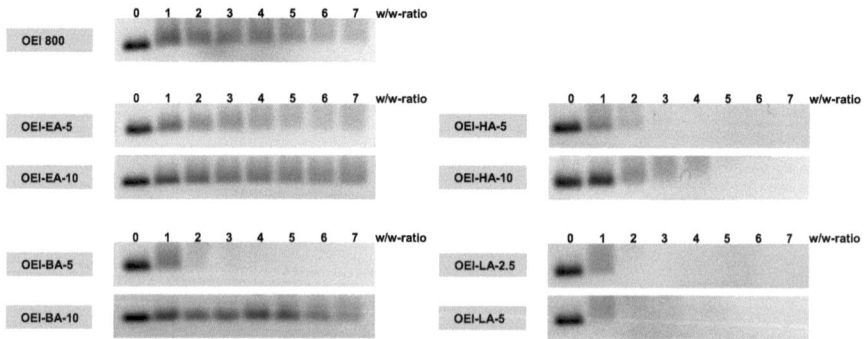

Figure 28: Gel shift assay for polyplexes of siRNA with different modified OEIs. The numbers represent polymer/siRNA mixing ratios (w/w).

3 RESULTS

While EtBr exclusion assay demonstrated, that unmodified OEI 800 and practically all modified oligoamines were effective in binding of siRNA at low ionic strength buffer HBG (data not shown), agarose gel shift assay illustrates differences in binding stability of the oligoamines to siRNA.

Unmodified OEI 800 showed relatively poor electrophoretic retention of siRNA in salt containing buffer and also the incorporation of short alkyl-chains (EA series) was not able to improve this property, presumably due to steric barriers between the charges of siRNA and oligoamine introduced by the bulky ethyl-acrylate groups. Polyplexes formed with OEI-BA-5 showed much higher complexation stability, whereas higher modification degree resulted in loss of binding ability. A similar trend could also be observed for modifications with higher hydrophobicity (HA series). Nevertheless, due to stronger hydrophobic interactions this effect is generally suppressed for longer alkyl-chains (LA series) featuring strong binding affinity to siRNA.

Moreover, co-formulations of OEI-HA-10 with OEI-LA-5 further increased the stabilization of polyplexes, whereas the incorporation of unmodified OEI 800 to the polyplexes resulted in decreased stability against dissociation as shown in Figure 29.

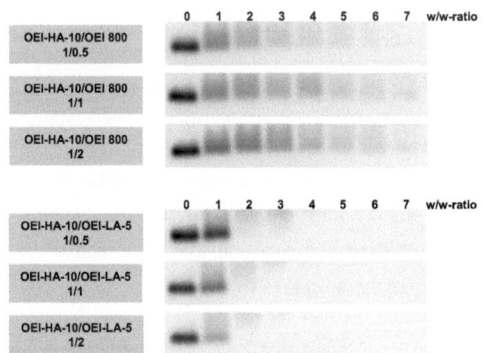

Figure 29: Gel shift assay for polyplexes of siRNA with co-formulation mixtures of OEI-HA-10/OEI 800 (upper panel) and OEI-HA-10/OEI-LA-5 (bottom panel). The numbers represent polymer/siRNA mixing ratios (w/w).

3.2.3.3 Colloidal stability of polyplex particles

Particle sizes of polyplexes prepared in HBG were determined using by photon correlation spectroscopy. Generally, polyplexes with OEI 800 were relatively small in size immediately after mixing in HBG. However, due to extremely low colloidal stability, rapid aggregation occurred within several minutes and, as a result, no particles were found 30min after

3 RESULTS

complexation, probably due to fast sedimentation of aggregates. The same behaviour was found in case of EA and BA modifications as shown in Table 6.

polymer	w/w	mean diameter [nm]	zeta-potential [mV]
OEI-EA-5	2/1	n.a.	7.2 ± 0.7
	4/1	n.a.	10.6 ± 1.0
OEI-EA-10	2/1	n.a.	-4.4 ± 0.6
	4/1	n.a.	-1.0 ± 0.3
OEI-BA-5	2/1	902 ± 15	16.5 ± 0.2
	4/1	n.a.	7.9 ± 1.7
OEI-BA-10	2/1	n.a.	-30.6 ± 3.7
	4/1	n.a.	-29.1 ± 0.8
OEI-HA-5	2/1	580 ± 10	12.9 ± 0.2
	4/1	208 ± 2	22.9 ± 0.4
OEI-HA-10	2/1	171 ± 2	21.0 ± 0.3
	4/1	194 ± 3	25.0 ± 1.8
OEI-LA-2,5	2/1	238 ± 4	8.4 ± 0.8
	4/1	136 ± 4	29.2 ± 1.0
OEI-LA-5	2/1	137 ± 2	40.9 ± 1.2
	4/1	119 ± 2	42.3 ± 1.6

n.a. not available due to low count rate

Table 6: Particle size and zeta-potential of polyplex particles for siRNA formulations with modified OEIs. Hydrodynamic diameters of complexes were determined in HBG by dynamic light scattering.

Only in case of HA and LA modifications, polyplexes were small in size and showed good colloidal stability even 4h after complexation as shown in Table 7. Apparently hydrophobic interactions between oligoamines with long hydrophobic alkyl-chains (HA and LA) could provide additional binding of excess oligoamine to the surface of the particle and, thus, provide an improved stability of the polyplexes.

polymer	w/w	mean diameter [nm]				zeta-potential [mV]			
		t = 0	t = 2h	t = 4h	t = 24h	t = 0	t = 2h	t = 4h	t = 24h
OEI-HA-10	2/1	223 ± 4	292 ± 14	272 ± 9	638 ± 17	18.0 ± 0.6	18.1 ± 0.9	17.9 ± 0.5	12.3 ± 0.2
	4/1	119 ± 1	116 ± 1	152 ± 1	229 ± 7	24.2 ± 1.2	26.4 ± 3.4	28.0 ± 0.4	20.5 ± 1.1
OEI-LA-5	2/1	122 ± 1	243 ± 4	168 ± 3	139 ± 2	36.7 ± 1.0	38.5 ± 2.4	42.1 ± 1.2	26.4 ± 1.9
	4/1	125 ± 4	197 ± 2	129 ± 1	148 ± 1	40.8 ± 0.8	40.6 ± 0.8	35.9 ± 1.2	31.0 ± 0.8
OEI-HA-10/OEI-LA-5 1/0,5	2/1	162 ± 2	215 ± 3	157 ± 2	183 ± 3	33.6 ± 1.3	29.3 ± 0.5	28.4 ± 2.5	25.6 ± 1.0
	4/1	137 ± 1	125 ± 1	108 ± 3	174 ± 2	34.6 ± 1.5	30.8 ± 3.0	31.4 ± 1.2	33.2 ± 1.5
OEI-HA-10/OEI-LA-5 1/1	2/1	154 ± 5	132 ± 2	185 ± 4	145 ± 2	30.6 ± 1.2	26.2 ± 1.6	34.4 ± 0.6	28.9 ± 0.9
	4/1	128 ± 3	133 ± 1	176 ± 3	139 ± 1	41.1 ± 0.2	39.1 ± 1.9	35.8 ± 2.0	34.0 ± 3.7
OEI-HA-10/OEI-LA-5 1/2	2/1	131 ± 1	129 ± 1	137 ± 1	151 ± 1	38.5 ± 0.6	34.8 ± 2.2	38.0 ± 3.3	37.3 ± 1.4
	4/1	145 ± 4	134 ± 1	194 ± 3	145 ± 4	40.6 ± 2.1	36.5 ± 1.6	39.6 ± 0.6	39.1 ± 1.6

Table 7: Colloidal stability of polyplex particles for siRNA formulations with OEI-HA-10, OEI-LA-5 and OEI-HA-10/OEI-LA-5. Hydrodynamic diameters of complexes were determined in HBG by dynamic light scattering.

3 RESULTS

Co-formulation of OEI-HA-10 with OEI-LA-5 shows additional advantage of colloidal stabilization, which could be potentially important regarding stability in the blood stream in view of in vivo applications.

3.2.3.4 Influence of conjugates on cytotoxicity

To investigate the influence of surface modification on cytotoxicity, the metabolic activity of cells was monitored after treatment with various concentrations of plain polymers as shown in Figure 30.

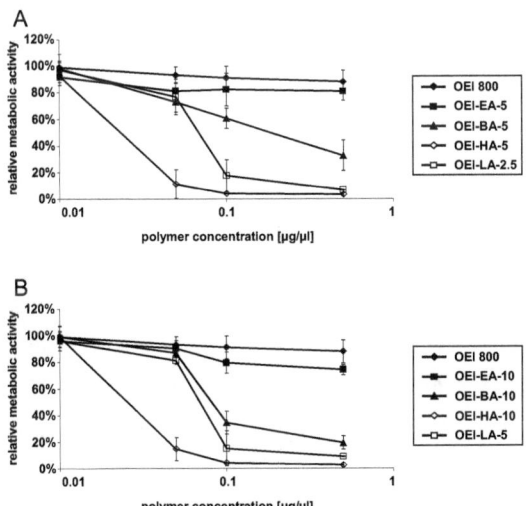

Figure 30: Cell viability of Neuro2A/eGFPLuc cells monitored by MTT assay as a function of polymer concentration. (A) OEIs with lower modification degree, (B) OEIs with higher modification degree.

Cytotoxicity studies generally revealed that increasing hydrophobicity of oligoamines results in decreased metabolic activity. However, the LA series, representing the highest hydrophobic modification, did not exhibit such a decreased metabolic activity compared to the HA series. Notably, the toxicity of hexyl-acrylate modified OEI is relatively high in comparison to all other conjugates, regarding both lower and higher modification degrees. Increased hydrophobicity of the HA series obviously results in strong interactions with lipid membranes, which may promote transfer across cellular barriers, e.g. by lysis of endosomes, but also lead to high toxicity during in vivo applications and, thus, may represent a problem.

Nevertheless, the acrylate ester bonds enable fast degradation of conjugates resulting in far less toxic components as shown in Figure 31.

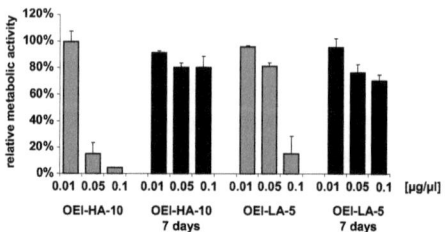

Figure 31: Cell viability of Neuro2A/eGFPLuc cells monitored by MTT assay before and after degradation of modified oligoamines as a function of polymer concentration. Grey bars indicate oligoamines without preincubation, black bars indicate oligoamines after preincubation at 37°C for 7 days in HBG.

The incubation of OEI-HA-10 and OEI-LA-5 at 37°C for 7 days in HBG led to a remarkable decrease of toxicity, presumably due to degradation of the ester bonds, which was further confirmed by ^{1}H-NMR spectroscopy (see Figure 27). Thus, long-term toxicity does not seem to be problematic.

3.2.3.5 siRNA delivery efficiency: structure-activity relationship

To evaluate the siRNA delivery efficiency of modified polymers, reporter gene silencing studies were carried out using Neuro2A/eGFPLuc, murine neuroblastoma, HUH7/eGFPLuc, human hepatocellular carcinoma and H1299/Luc, human lung carcinoma cell lines, stably expressing the luciferase reporter gene. Polyplexes were prepared in HBG containing either LucsiRNA targeting the firefly luciferase or siCONTROL as non-targeting control siRNA to clearly distinguish between specific gene silencing and unspecific toxic side effects due to the carrier system. Figure 32 shows the siRNA gene silencing capability of all modified conjugates 48h after initial siRNA delivery on Neuro2A/eGFPLuc cells in serum containing (10% FCS) growth medium.

3 RESULTS

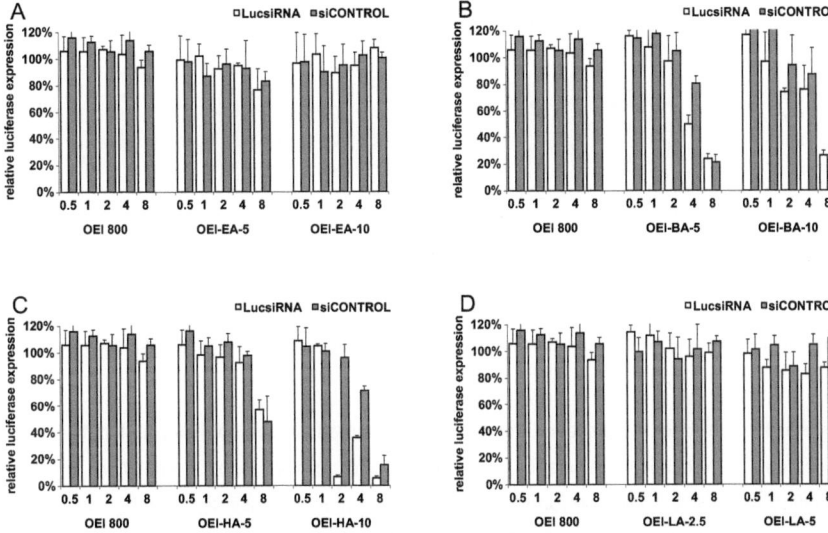

Figure 32: siRNA gene silencing efficiency on Neuro2A/eGFPLuc cells using 500ng siRNA (equal to 380nM). (A) OEI-EA series, (B) OEI-BA series, (C) OEI-HA series, (D) OEI-LA series. White bars indicate complexes containing luciferase siRNA, grey bars indicate complexes containing control siRNA. The numbers on the x-axis represent polymer/siRNA mixing ratios (w/w).

Only OEI-HA-10 formulations were found to be promising siRNA carriers effective in knockdown of luciferase expression without unspecific toxicity. Even formulations with a lower degree of modification (OEI-HA-5) were completely ineffective for siRNA delivery. Additionally, neither formulations with shorter hydrophobic chains (EA and BA series) nor formulations with longer hydrophobic chains (LA series) showed any reduction of luciferase activity. Thus, OEI-HA-10 seems to have an optimal structure with enhanced endosomolytic properties, arising from 10 hexyl-acrylate residues per OEI molecule, which was found to have an optimal knockdown efficiency at a w/w ratio of 2/1 as shown in Figure 33.

3 RESULTS

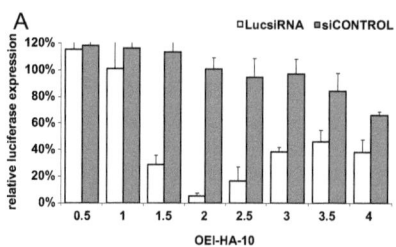 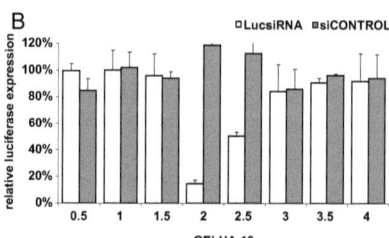

Figure 33: siRNA gene silencing efficiency of OEI-HA-10 on Neuro2A/eGFPLuc cells using 500ng siRNA (equal to 380nM). (A) 10% serum present in medium, **(B)** 50% serum present in medium. White bars indicate complexes containing luciferase siRNA, grey bars indicate complexes containing control siRNA. The numbers on the x-axis represent polymer/siRNA mixing ratios (w/w).

Interestingly, further increase of the OEI-HA-10 content in the formulation led to certain decrease of knockdown efficiency, while unspecific toxicity of the formulation remained unaffected (Figure 33A). The formulations were also able to cause significant knockdown of luciferase expression even in 50% serum containing medium, however only at w/w ratios of 2/1 – 2.5/1 (Figure 33B).

Additionally, OEI-HA-10 was tested for siRNA delivery on HUH7/eGFPLuc and H1299/Luc cells as shown in Figure 34.

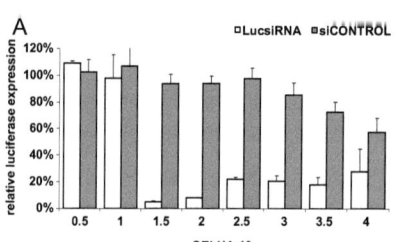 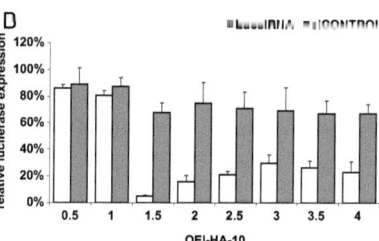

Figure 34: siRNA gene silencing efficiency of OEI-HA-10 on HUH7/eGFPLuc cells (A) and H1299/Luc cells (B) using 500ng siRNA (equal to 380nM). In case of H1299/Luc cells medium change was performed 24h after siRNA delivery. White bars indicate complexes containing luciferase siRNA, grey bars indicate complexes containing control siRNA. The numbers on the x-axis represent polymer/siRNA mixing ratios (w/w).

A similar knockdown efficiency profile for OEI-HA-10 could be observed again on both cell lines. Such unusual bell-shaped behavior cannot be explained by formation of polyplex aggregates, which (in the case of standard PEI 22) may mediate more efficient delivery of nucleics acid in comparison to small particles[284]. OEI-HA-10 polyplex particles, however, revealed small particle sizes at all ratios under investigation (Table 6).

3.2.3.6 Lytic activity of conjugates

Similar to several endosomolytic peptides, enhancement of hydrophobicity in the oligoamine structure could increase the ability of conjugates to lyse lipid membranes and, thus, facilitate efficient delivery of siRNA into the cytoplasm. To evaluate the membrane destabilizing activity of hydrophobic modified oligoamines, lytic activites of the conjugates were investigated in an erythrocyte leakage assay as shown in Figure 35.

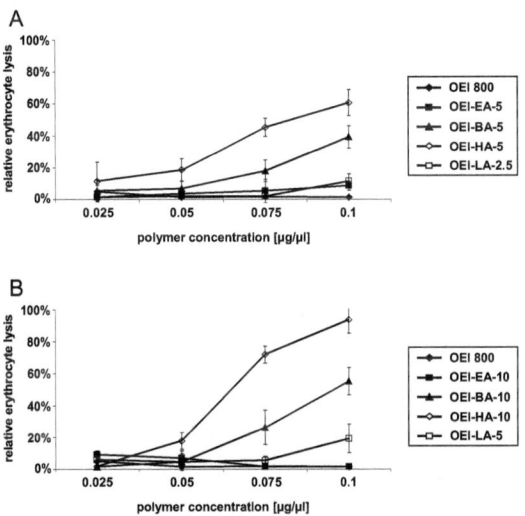

Figure 35: Haemolytic activity of plain conjugates. (A) OEIs with lower modification degree, (B) OEIs with higher modification degree. Erythrocytes at a concentration of 4% (V/V) (~ approximately 10^8 erythrocytes per ml) were incubated with increasing concentrations of conjugates in HBG containing 10% FCS at 37°C for 45min. Haemolysis was determined by UV measurement at 405nm relative to Triton X (100% lysis).

In order to adjust the experimental conditions to in vivo situation, haemolytic activity of plain conjugates was studied in the presence of 10% serum.

Modifications with acrylates of lowest hydrophobicity (EA series) and highest hydrophobicity (LA series) resulted in very low haemolytic activity. In contrast, in case of BA and HA modifications, significant lytic properties of the conjugates were observed discovering similar toxicity profiles as found in the cytotoxicity and gene silencing studies.

Hydrophobic OEI-HA-10 exposed highest haemolytic activity (> 90% of hemoglobin release) exhibiting distinct membrane interactions and finally cell lysis. Thus, the addition of hexyl chains into OEI strongly increased the membrane destabilization activity of OEI which seems to be responsible for the high toxicity of the conjugate but also for effective intracellular transport across cellular membranes, presumably by promoting escape from the endosomes.

3.2.3.7 Co-formulation with helper polymers and lipids for improved siRNA delivery

Incorporation of active transfection agents into an appropriate formulation could often improve physicochemical parameters (e.g. particle size, stability) and biological parameters (e.g. efficiency, biocompatibility) of the final formulation. Helper polymers[285] or helper lipids[286-290] could be utilized for this purpose.

Due to the strong lytic activity and cytotoxicity of OEI-HA-10 formulations, co-formulations with different hydrophilic and hydrophobic agents (such as OEI 800, OEI-LA-5, DOPE, DOPC and DPPC) were investigated in order to improve the biocompatibility and efficiency of the formulations in vitro.

A promising optimization procedure was a dilution of OEI-HA-10 in the formulation by the far less toxic but ineffective oligoamines OEI 800 and OEI-LA-5, respectively, as shown in Figure 36.

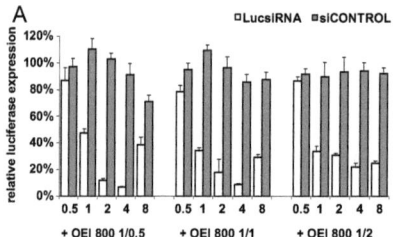
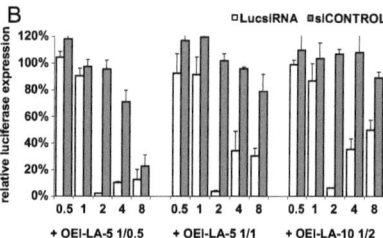

Figure 36: siRNA gene silencing efficiency on Neuro2A/eGFPLuc cells using 500ng siRNA (equal to 380nM). (A) OEI-HA-10/OEI 800 series, (B) OEI-HA-10/OEI-LA-5 series. White bars indicate complexes containing luciferase siRNA, grey bars indicate complexes containing control siRNA. The numbers on the x-axis represent total polymer/siRNA mixing ratios (w/w).

Substitution of a certain amount of OEI-HA-10 by relatively non-toxic low molecular weight OEI 800 or OEI-LA-5 oligoamines in the formulation led to a remarkable decrease of toxicity associated with a gradual decrease of the OEI-HA-10 content.

Additionally, the efficiency profile was further improved, as for example OEI-HA-10/OEI 1/2 (w/w) formulations with siRNA induced efficient reporter gene silencing at lower mixing ratios in comparison to pure OEI-HA-10 revealing that efficiency was less dependent on the mixing ratio. The combination with OEI-LA-5 shows further advantage of colloidal stabilization, which could be potentially important in view of in vivo applications, while the incorporation of OEI 800 to the formulation decreases colloidal stability (see Figure 29, Table 7).

Importantly, unmodified OEI 800 and OEI-LA-5 formulations alone were not active in siRNA delivery at all studied mixing ratios and serve only as helper reagents in the formulation.

Another approach was co-formulation of OEI-HA-10 with different helper lipids, such as DOPE, DOPC and DPPC, as shown in Figure 37.

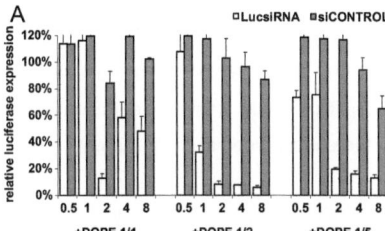
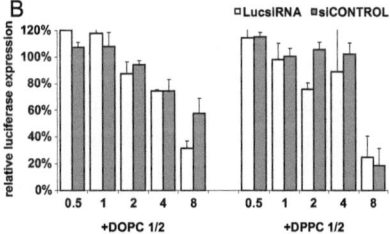

Figure 37: siRNA gene silencing efficiency on Neuro2A/eGFPLuc cells using 500ng siRNA (equal to 380nM). (A) OEI-HA-10/DOPE series, (B) OEI-HA-10/DOPC and OEI-HA-10/DPPC series. White bars indicate complexes containing luciferase siRNA, grey bars indicate complexes containing control siRNA. The numbers on the x-axis represent total polymer/siRNA mixing ratios (w/w).

The hydrophobic structure of OEI-HA-10 allows interaction with the hydrophobic domains of the helper lipids which results in decreased toxicity of the formulations.

Combinations of OEI-HA-10 with DOPE resulted in only slightly less toxic formulations compared to OEI-HA-10 alone, however the efficiency profile was greatly improved (Figure 37A). Particularly, formulations with lower amounts of DOPE showed a similar efficiency profile as pure OEI-HA-10, while formulations with higher amounts of DOPE resulted in efficient knockdown effect over a broad range of w/w ratios. The endosomolytic character of the DOPE lipid seems to be the reason for improved efficiency of OEI-HA-10/DOPE formulations, as it is known to destabilize the bilayer structure after incorporation into lipid membranes due to its inverse head-tail symmetry.

Formulations with DPPC and DOPC could also decrease the toxicity but also dramatically decreased the efficiency of the formulations (Figure 37B). Moreover, substitution of a certain amount of DPPC or DOPC with DOPE was also not able to improve the gene silencing efficiency (data not shown).

3.3 Bioresponsive endosomolytic conjugates for siRNA delivery

Poor endosomal release after cellular internalization is one of the major barriers for efficient nucleic acid delivery. Ways to ensure effective release of the payload from the endosome into the cytoplasm are for example: making use of the "proton sponge" effect of some polycations; and/or incorporation of lytic lipid moieties or membrane active peptides into the polymer. Thus, several endosomolytic peptides, which were derived from viruses[224,232], toxins[233,269,291] or synthetically designed[235] had been applied for such purpose[225,237,240,265,292].

In particular, the membrane active peptide melittin, a 26 amino acid peptide, was incorporated into polycations in order to promote quick escape from the endosome after cellular uptake avoiding subsequent degradation in the endolysosomal compartment.

3.3.1 DMMAn-Mel modified conjugates for siRNA delivery

3.3.1.1 Design of endosomolytic conjugates

In an approach to generate siRNA carrier systems with enhanced endosomal escape properties, Martin Meyer (PhD thesis 2009, LMU) functionalized polycations such as PLL and PEI 25 with PEG and a pH-responsive endosomolytic melittin peptide.

As pure melittin displays a strong lytic activity, which is quite unfavorable in the extracellular environment resulting in toxic side effects, the amines of melittin were modified with pH labile dimethylmaleic anhydride, which minimizes the lytic activity at extracellular neutral pH. After endosomal acidification the DMMAn protecting groups are cleaved and lytic activity is restored[265-266]. Thus, endosomal acidification is exploited for a triggered activation of the lytic activity in the intracellular compartment. The design of the pH-resonsive melittin conjugate is shown schematically in Figure 38.

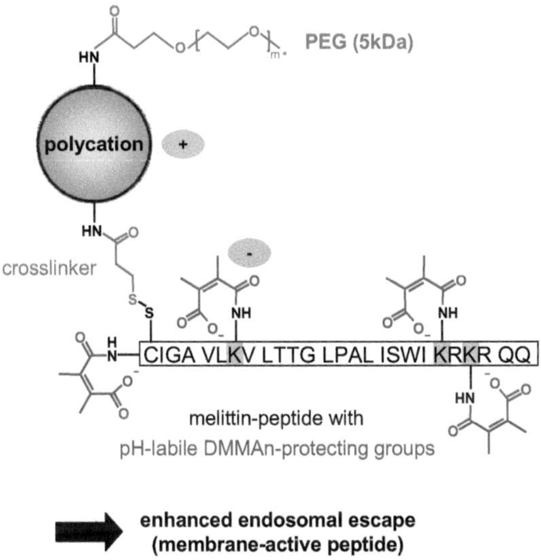

3 RESULTS

Figure 38: Design of the pH-responsive melittin conjugate. Endosomal acidification triggers fast cleavage of the dimethylmaleic anhydride protecting groups and restores lytic activity of melittin. Conjugate synthesis and further characterization was performed by Martin Meyer as part of his PhD thesis (LMU, 2009).

In order to synthesize conjugates on the basis of a reversibly acylated melittin with the favored bioresponsive lytic activity, DMMAn-Mel was covalently attached via the N-terminal cysteine residue to the PDP-modified polycations PLL, PEI 25 and OEI-HD-1 by disulfide bond formation. Melittin was linked via its N-terminus which also reduces the cytotoxic potential at neutral pH[328].

Poly-L-lysine (MW 32kDa), one of the first polycations used for polyplex formation, represents a biodegradable polycation due to its peptidic nature. Furthermore, a high molecular weight PLL guarantees an overall positive charge of the conjugate which is necessary for cell interaction and internalization, however subsequent escape from the endosomes into the cytoplasm presents a major bottleneck, which results in low nucleic acid delivery efficiency. In contrast, branched PEI (MW 25kDa), currently one of the most frequently used polycations, may promote its escape to the cytosol from endosomes via the "proton sponge" effect, however it still remains ineffective for siRNA delivery.

Since PEI 25 represents a non-degradable high molecular weight polyethylenimine, the biodegradable oligoethylenimine derived polymer OEI-HD-1 was additionally chosen as backbone for modification with DMMAn-Mel.

Hydrophilic polyethylene glycol (MW 5kDa) was grafted onto the polycations prior to peptide attachment in order to prevent aggregation after mixing with nucleic acids and to enhance solubility and stability of nucleic acid complex formation. The final conjugates had a molar ratio of approximately 1/1/8 (PEG/polycation/DMMAn-Mel).

3.3.1.2 siRNA binding ability

The capability of polymers to condense siRNA, in order to form complexes suitable for cell entry, was studied using an EtBr exclusion assay. The reduction of relative fluorescence was measured as a function of increasing polymer/siRNA mixing ratios as shown in Figure 39.

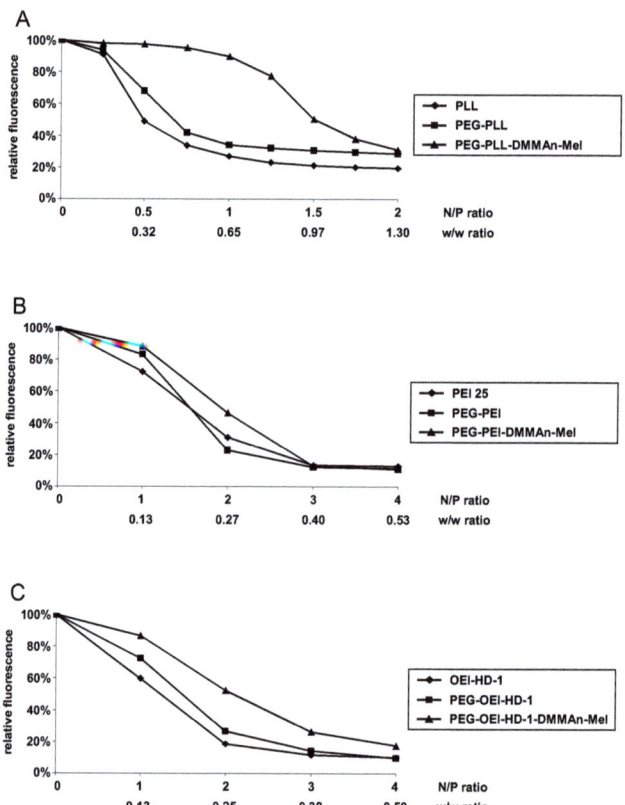

Figure 39: siRNA binding affinity of conjugates determined by EtBr exclusion assay. (A) PLL based conjugates, (B) PEI based conjugates, (C) OEI-HD-1 based conjugates. The numbers on the x-axis represent polymer/siRNA mixing ratios (N/P or w/w). Data for PLL and PEI based conjugates were generated by Martin Meyer as part of his PhD thesis (LMU 2009).

3 RESULTS

Investigating the influence of peptide modification on siRNA binding, all polymers were found to be effective in binding of siRNA at low ionic strength buffer HBG. The effect of PEGylation on siRNA binding ability was negligible, however conjugates modified with negatively charged DMMAn-Mel peptides required slightly higher amounts of polycation to achieve complete binding of siRNA compared to their unmodified counterparts.

Moreover, siRNA polyplexes prepared at w/w ratios of 1/1 and 2/1 in HBG showed small particle sizes of about 15 - 40nm in case of all conjugates indicating almost monomolecular structures, measured by fluorescence correlation spectroscopy using Cy5 labeled siRNA[247].

3.3.1.3 siRNA delivery efficiency and toxicity of conjugates

To evaluate the influence of endosomolytic peptides on siRNA delivery efficiency, reporter gene silencing studies were carried out using Neuro2A/eGFPLuc, murine neuroblastoma and HUH7/eGFPLuc, human hepatocellular carcinoma cell lines, stably expressing the luciferase reporter gene. Polyplexes were prepared in HBG containing either LucsiRNA targeting the firefly luciferase or siCONTROL as non-targeting control siRNA to clearly distinguish between specific gene silencing and unspecific toxic side effects due to the carrier system.

3.3.1.3.1 PEG-PLL conjugates

Figure 40A shows the siRNA gene silencing capability of PLL based conjugates 48h after initial siRNA delivery on Neuro2A/eGFPLuc cells in serum containing (10% FCS) growth medium.

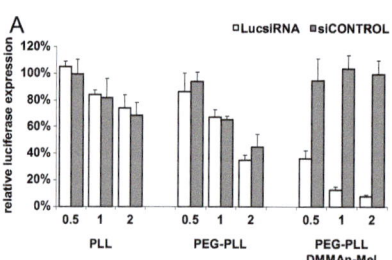 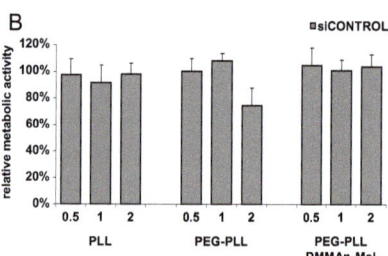

Figure 40: siRNA gene silencing efficiency of PLL based conjugates on Neuro2A/eGFPLuc cells (A) and cell viability of Neuro2A/eGFPLuc cells monitored by MTT assay (B) using 500ng siRNA (equal to 380nM). White bars indicate complexes containing luciferase siRNA, grey bars indicate complexes containing control siRNA. The numbers on the x-axis represent polymer/siRNA mixing ratios (w/w).

siRNA formulations with PLL or PEG-PLL were not able to induce any significant reduction in luciferase expression without unspecific toxicity. But after modification with DMMAn-Mel, siRNA delivery efficiency was greatly enhanced as demonstrated by excellent reporter gene knockdown (> 90%). Notably, PEG-PLL polyplexes displayed slightly increased toxicity at higher w/w ratios contrary to the other conjugates, as possibly in this particular case, PEGylation (unless applied with a higher molecular weight or a higher quantity) might provide detergent-like properties as copolymer with the polycation. Nevertheless, conjugation with DMMAn-Mel considerably reduced the acute toxicity of the polycation PEG-PLL.

Figure 40B shows the influence on metabolic activity of Neuro2A/eGFPLuc cells after polyplex treatment under same conditions, which is consistent with the unspecific reduction of luciferase expression induced by siCONTROL formulations.

Even at lower amounts of siRNA, PEG-PLL-DMMAn-Mel achieved significant knockdown of luciferase expression, if the polymer/siRNA mixing ratio was kept constant at 2/1 as shown in Figure 41.

3 RESULTS

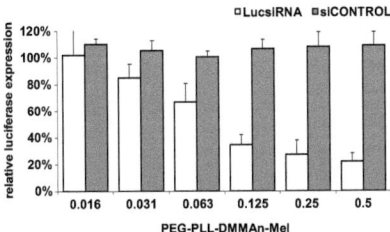

Figure 41: siRNA gene silencing efficiency of PEG-PLL-DMMAn-Mel on Neuro2A/eGFPLuc cells using 16ng - 500ng siRNA (equal to 12nM - 380nM). The concentration of unmodified PEI 25 was kept constant at 2.5µg/ml, the concentration of modified PEIs was kept constant at 20µg/ml. White bars indicate complexes containing luciferase siRNA, grey bars indicate complexes containing control siRNA. The numbers on the x-axis represent the amount of applied siRNA [µg].

3.3.1.3.2 PEG-PEI conjugates

Figure 42A shows the siRNA gene silencing capability of PEI based conjugates 48h after initial siRNA delivery on Neuro2A/eGFPLuc cells in serum containing (10% FCS) growth medium.

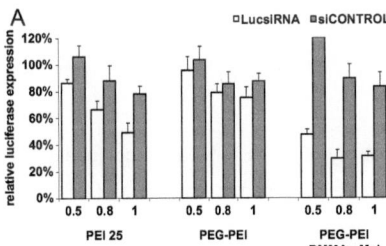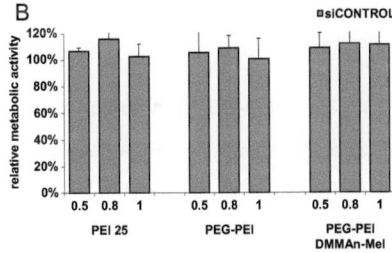

Figure 42: siRNA gene silencing efficiency of PEI based conjugates on Neuro2A/eGFPLuc cells (A) and cell viability of Neuro2A/eGFPLuc cells monitored by MTT assay (B) using 500ng siRNA (equal to 380nM). White bars indicate complexes containing luciferase siRNA, grey bars indicate complexes containing control siRNA. The numbers on the x-axis represent polymer/siRNA mixing ratios (w/w).

As expected, siRNA formulations with PEI 25 or PEG-PEI were not able to induce any significant reduction in luciferase expression without unspecific toxicity. Modification with DMMAn-Mel, however, resulted again in strongly enhanced gene silencing efficiency (> 70%).

Figure 42B shows the influence on metabolic activity of Neuro2A/eGFPLuc cells after polyplex treatment under same conditions, which is consistent with the unspecific reduction of luciferase expression (siCONTROL formulations).

PEG-PEI-DMMAn-Mel, was additionally tested for siRNA delivery on HUH7/eGFPLuc cells as shown in Figure 43.

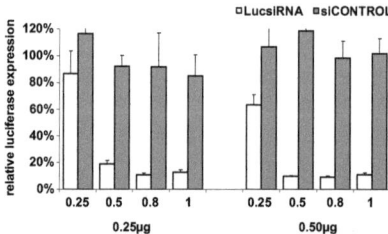

Figure 43: siRNA gene silencing efficiency of PEG-PEI-DMMAn-Mel on HUH7/eGFPLuc cells using 250ng or 500ng siRNA (equal to 190nM or 380nM). White bars indicate complexes containing luciferase siRNA, grey bars indicate complexes containing control siRNA. The numbers on the x-axis represent polymer/siRNA mixing ratios (w/w).

Also on this cell line, PEG-PEI-DMMAn-Mel showed greatly enhanced siRNA delivery efficiency demonstrated by excellent marker gene knockdown. Moreover, the knockdown effect was more pronounced on this cell line and remained constant even at lower amounts of siRNA.

3.3.1.3.3 PEG-OEI-HD-1 conjugates

Figure 44A shows the siRNA gene silencing capability of OEI-HD-1 based conjugates 48h after initial siRNA delivery on Neuro2A/eGFPLuc cells in serum containing (10% FCS) growth medium.

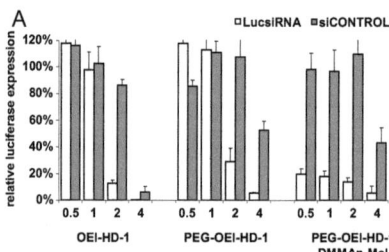 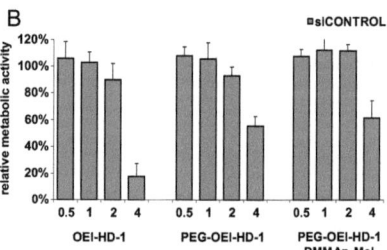

Figure 44: siRNA gene silencing efficiency of OEI-HD-1 based conjugates on Neuro2A/eGFPLuc cells (A) and cell viability of Neuro2A/eGFPLuc cells monitored by MTT assay (B) using 250ng siRNA (equal to 190nM). White bars indicate complexes containing luciferase siRNA, grey bars indicate complexes containing control siRNA. The numbers on the x-axis represent polymer/siRNA mixing ratios (w/w).

3 RESULTS

The intrinsic siRNA delivery activity of the biodegradable polymer OEI-HD-1 was only marginally affected by PEGylation, as the optimal knockdown effect was marginally shifted to higher w/w ratios associated with slightly reduced toxicity. Furthermore, conjugation of DMMAn-Mel to PEG-OEI-HD-1 greatly enhanced the siRNA delivery activity as demonstrated by excellent knockdown of luciferase expression associated with reduced toxicity of the conjugate. In addition, knockdown efficiency was even shifted to lower w/w ratios converting the formulation into a highly efficient siRNA carrier in vitro.

Figure 44B shows the influence on metabolic activity of Neuro2A/eGFPLuc cells after polyplex treatment under same conditions, which is consistent with the unspecific reduction of luciferase expression induced by siCONTROL formulations.

3.3.1.4. pH triggered lytic activity of conjugates

To evaluate the desired pH dependant membrane destabilizing activity of DMMAn modified melittin, lytic activites of the conjugates were investigated in an erythrocyte leakage assay before and after preincubation at endosomal pH.

Figure 45 shows OEI-HD-1 based conjugates as representative example for the triggered lytic activity of DMMAn-Mel, which is similar to that of PLL and PEI based conjugates (data not shown).

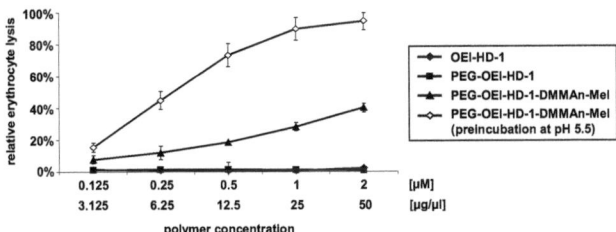

Figure 45: Haemolytic activity of plain conjugates. Erythrocytes at a concentration of 0.1% (V/V) (~ approximately 10^7 erythrocytes per ml) were incubated with increasing concentrations of conjugates in HBG at 37°C for 30min. PEG-OEI-HD-1-DMMAn-Mel conjugates were applied directly or after acidic preincubation at pH 5.5 for 30min at room temperature. Haemolysis was determined by UV measurement at 405nm relative to Triton X (100% lysis).

While OEI-HD-1 and PEG-OEI-HD-1 did not show any haemoglobin release from treated erythrocytes, DMMAn-Mel modified conjugates exhibited slight membrane damage in a dose dependent manner. In contrast, lytic activity of PEG-OEI-HD-1-DMMAn-Mel was greatly enhanced (> 90% haemoglobin release) after acidic preincubation at pH 5.5 due to pH-specific cleavage of the DMMAn protecting groups.

3.3.2 Covalently attached siRNA-polymer conjugates for improved siRNA delivery

3.3.2.3 Design of dynamic functionalized siRNA carriers

Extracellular stability of electrostatically formed siRNA polyplexes is a significant concern in the delivery process. Especially in vivo delivery entails the danger that negatively charged molecules, e.g. serum proteins or the extracellular matrix, can disrupt such complexes, which results in complete disassembly of polyplexes before reaching the target site of action[157-158]. On the other hand, after specific cellular uptake the same carrier system should provide efficient release of the nucleic acid into the cytosol. As the described delivery functions are required at different time points of the delivery process, siRNA carriers have to be dynamic in their characteristics, like natural viruses, to be most effective at the different steps of extracellular and intracellular delivery[242,293].

Consequently, to overcome the risk of polyplex dissociation in the extracellular environment, siRNA was additional covalently attached to the already well established pH responsive PEG-PLL-DMMAn-Mel conjugate. Thereby the requested additional changes were programmed into the carrier system by introduction of bioresponsive cleavable bonds depending on the reductive cleavage of disulfides in the cytoplasm. A schematic structure of the bioresponsive siRNA-polymer conjugate is shown in Figure 46.

3 RESULTS

A

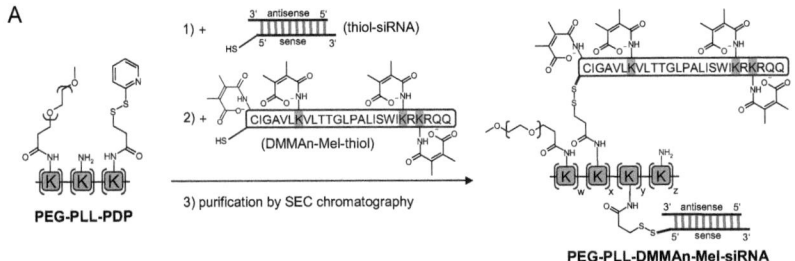

B

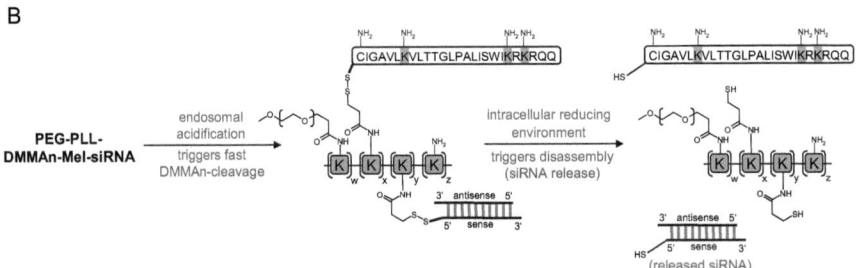

Figure 46: Design of the bioresponsive PEG-PLL-DMMAn-Mel-siRNA conjugate (A), pH triggered cleavage of DMMAn protecting groups after endosomal acidification and release of siRNA upon disulfide cleavage in reducing environment (B). [K] represents L-lysine monomers of the PLL backbone. Conjugate synthesis and further characterization was performed by Martin Meyer as part of his PhD thesis (LMU, 2009).

Synthesis of the PEG-PLL-DMMAn-Mel-siRNA conjugate was carried out similarly to the PEG-PLL-DMMAn-Mel synthesis, however in high salt concentration of 1.5M NaCl to prevent aggregation of anionic siRNA and DMMAn-Mel with the cationic PEG-PLL-PDP[85]. First, siRNA was covalently coupled to PEG-PLL-PDP by bioreducible disulfide bonds as linkage between the 5' end of the sense strand of thiol modified siRNA and PDP modified conjugate. Secondly, thiol containing DMMAn-Mel was coupled to the remaining PDP groups followed by purification (Figure 46A).

At physiological pH the amines of melittin were masked with dimethylmaleic anhydride, which minimizes the lytic activity of the conjugate. Upon endosomal acidification, however, DMMAn protecting groups are cleaved from the conjugated melittin peptides and lytic activity is restored promoting intracellular release out of the endosomes. After reaching the cytoplasm, the intracellular reductive environment triggers cleavage of the disulfide bonds enabling release of siRNA from the conjugate and subsequent RISC activation (Figure 46B).

The final luciferase siRNA conjugates had a molar ratio of approximately 1/1/7.5/1.5 (PEG/PLL/DMMAn-Mel/siRNA), the final control siRNA conjugates had a molar ratio of approximately 1/1/6/1.3 (PEG/PLL/DMMAn-Mel/siRNA). This corresponds to a PLL/siRNA w/w ratio of about 1.6/1, close to the optimal ratio found in the gene silencing experiments for the corresponding siRNA polyplexes.

3.3.2.4 Particle size determination of conjugates and polyplexes

With respect to potential in vivo applicability, particle sizes formed by the covalent siRNA-polymer conjugates were determined using photon correlation spectroscopy. PEG-PLL-DMMAn-Mel-siRNA conjugates displayed particles in the range of 80 - 300nm strongly dependant on the salt concentration[85].

Transmission electron microscopy investigations further confirmed these results presenting particle sizes of approximately 30 - 50nm (width) × 65 - 100nm (length) for covalent siRNA-polymer conjugates. Similar particle sizes of approximately 40 - 50nm (width) × 60 - 100nm (length) were found for the corresponding analogous PEG-PLL-DMMAn-Mel/siRNA polyplexes (prepared at a w/w ratio of 2/1) as shown in Figure 47.

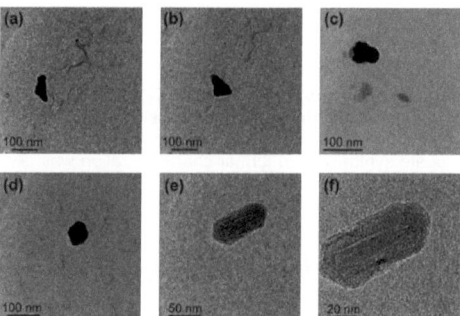

Figure 47: Transmission electron microscopy images showing the morphology of different siRNA particles. (a) - (c) presents PEG-PLL-DMMAn-Mel-siRNA conjugates, (d) - (f) presents PEG-PLL-DMMAn-Mel / siRNA polyplexes, (f) presents a higher magnified view of (e). Measurements were performed in cooperation with Dr. Daniel Kiener, research group Prof. C. Scheu, Department of Chemistry and Biochemistry (LMU).

3.3.2.5 siRNA delivery efficiency and toxicity of PEG-PLL-DMMAn-Mel-siRNA conjugates

To evaluate the siRNA delivery efficiency of the covalent siRNA-polymer conjugates, reporter gene silencing studies were carried out using a Neuro2A/eGFPLuc, murine neuroblastoma cell line, stably expressing the luciferase reporter gene. siRNA-polymer conjugates and the corresponding formulation containing electrostatically complexed siRNA, respectively, were prepared in HBG containing either LucsiRNA targeting the firefly luciferase or siCONTROL as non-targeting control siRNA to clearly distinguish between specific gene silencing and unspecific toxic side effects due to the carrier system. Figure 48A shows the siRNA gene silencing capability of PEG-PLL-DMMAn-Mel-siRNA conjugates compared to the corresponding analogous PEG-PLL-DMMAn-Mel/siRNA polyplexes 48h after initial siRNA delivery on Neuro2A/eGFPLuc cells in serum containing (10% FCS) growth medium.

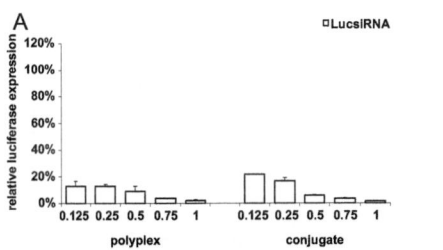
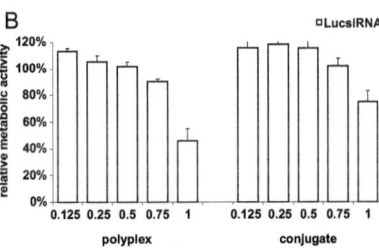

Figure 48: siRNA gene silencing efficiency on Neuro2A/eGFPLuc cells (A) and cell viability of Neuro2A/eGFPLuc cells monitored by MTT assay (B) using 125ng - 1000ng siRNA (equal to 95nM - 760nM). PEG-PLL-DMMAn-Mel-siRNA conjugates (right) were compared to the corresponding analogous PEG-PLL-DMMAn-Mel/siRNA polyplexes prepared at a w/w ratio of 2/1 (left). White bars indicate conjugates or complexes containing luciferase siRNA. The numbers on the x-axis represent the amount of applied siRNA [µg].

The biological efficiency of the covalent luciferase siRNA-polymer conjugate as well as the viability of treated cells was compared with the corresponding luciferase siRNA polyplex formulation, which was electrostatically complexed at a w/w ratio of 2/1.

High in vitro biocompatibility, i.e. absence of cytotoxicity, and efficient sequence-specific gene silencing (> 80%) was found for both formulations even at lower siRNA doses (0.125µg and 0.25µg) (Figure 48).

While cells maintained a high metabolic activity at siRNA doses from 0.125µg to 0.75µg, the metabolic activity was significantly reduced at the highest applied siRNA dose of 1µg (Figure 48B).

The PLL/siRNA w/w ratio of the covalent luciferase siRNA-polymer conjugates is 1.6/1 (molar ratio of PLL/siRNA is 0.68/1, charge ratio is about 2/1), which is quite similar to the polyplexes with a w/w ratio of 2/1 (molar ratio of PLL/siRNA is 0.85/1, charge ratio is about 3/1).

Additionally, the biological knockdown activity of the covalent luciferase siRNA-polymer conjugates was compared side-by-side with the covalent control siRNA-polymer conjugates over a larger range of siRNA concentrations as shown in Figure 49.

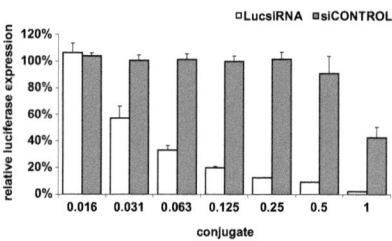

Figure 49: siRNA gene silencing efficiency of PEG-PLL-DMMAn-Mel-siRNA conjugates on Neuro2A/eGFPLuc cells using 15.6ng - 1000ng siRNA (equal to 12nM - 760nM). White bars indicate conjugates containing luciferase siRNA, grey bars indicate conjugates containing control siRNA. The numbers on the x-axis represent the amount of applied siRNA [µg].

Luciferase siRNA containing PEG-PLL-DMMAn-Mel-siRNA conjugates showed remarkable gene silencing effects even at low siRNA doses of 0.031µg, while no reduction of luciferase activity was observed with the control siRNA containing PEG-PLL-DMMAn-Mel-siRNA conjugates, unless the high (slightly cytotoxic) 1µg dose was applied (Figure 49).

3.3.2.6 pH triggered lytic activity of conjugates

To demonstrate the pH responsiveness of the conjugate, pH triggered lytic activity of PEG-PLL-DMMAn-Mel-siRNA conjugates was investigated in an erythrocyte leakage assay before and after preincubation at endosomal pH as shown in Figure 50.

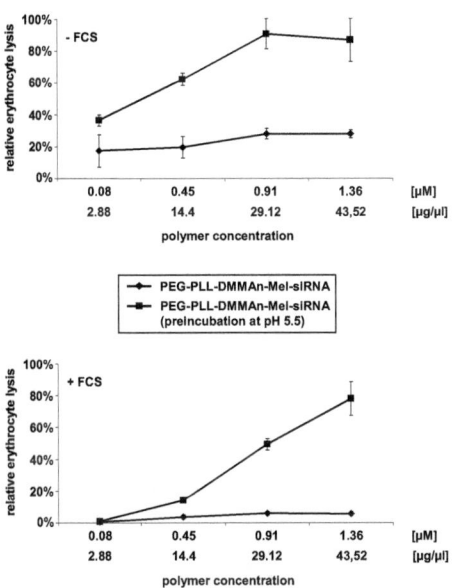

Figure 50: Haemolytic activity of plain conjugates. Erythrocytes at a concentration of 0.1% (V/V) (~ approximately 10^7 erythrocytes per ml) were incubated with increasing concentrations of conjugates in HBG (upper panel) or HBG containing 3% FCS (bottom panel) at 37°C for 10min (in case of HBG) or 20min (in case of HBG containing 3% FCS). PEG-PLL-DMMAn-Mel-siRNA conjugates were applied directly or after acidic preincubation at pH 5.5 for 30min at room temperature. Haemolysis was determined by UV measurement at 405nm relative to Triton X (100% lysis).

Erythrocytes were incubated with the covalent PEG-PLL-DMMAn-Mel-siRNA conjugates in HBG in the absence or presence of serum.

The lytic activity of conjugates was greatly enhanced after acidic preincubation at pH 5.5 (Figure 50), consistent with pH-specific cleavage of the DMMAn protecting groups (see Figure 46B).

This pH dependency was not negatively affected by the addition of serum (Figure 50, lower panel). In fact, haemolytic activity of DMMAn masked conjugates is strongly reduced by the presence of serum proteins, while the activity of unmasked conjugates after acidic preincubation at pH 5.5 is largely maintained (> 80% haemoglobin release).

3.3.2.7 Glutathione induced release of siRNA

To demonstrate the redox responsiveness of the covalent PEG-PLL-DMMAn-Mel-siRNA conjugate and to examine, if release of siRNA is possible in cells, the siRNA-polymer conjugate was incubated at 37°C with physiological glutathione (GSH) concentrations and release of siRNA was monitored by agarose gel electrophoresis as shown in Figure 51.

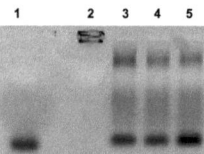

Figure 51: Gel shift assay for PEG-PLL-DMMAn-Mel-siRNA conjugates after glutathione treatment. Conjugates were incubated with 2 I.U. heparin (in order to eliminate electrostatic interactions between the polycation and the nucleic acid) and different amounts of glutathione at 37°C for 75min. Lane (1) 0.5µg siRNA + GSH 20mM, lane (2) siRNA conjugate (= 0.5µg siRNA), lane (3) siRNA conjugate (= 0.5µg siRNA) + GSH 1.25nM, lane (4) siRNA conjugate (= 0.5µg siRNA) + GSH 2.5nM, lane (5) siRNA conjugate (= 0.5µg siRNA) + GSH 5nM. Data were generated by Martin Meyer as part of his PhD thesis (LMU, 2009).

Glutathione is a natural reducing agent which is found in millimolar concentrations in the cytosol (1 - 10mM), whereas in the extracellular environment only micromolar concentrations are present[294].

From untreated siRNA-polymer conjugate no siRNA release is detected during agarose gel electrophoresis (Figure 51, lane 2). Moreover, neither glutathione (reducing agent) nor heparin (polyanion) treatment alone could induce a release of siRNA at conditions where the polyanion heparin causes disassembly of electrostatic siRNA polyplexes, indicating that siRNA is associated with PLL both covalently and electrostatically (data not shown).

Treatment with heparin under reducing conditions with different glutathione concentrations could efficiently trigger the release of siRNA from the PEG-PLL-DMMAn-Mel-siRNA conjugate (Figure 51, lanes 3 - 5).

4 DISCUSSION

Nucleic acid based therapy holds much promise in the treatment of many genetic and acquired diseases, although the clinical use is still limited by the lack of safe and efficient nucleic acid delivery systems. Synthetic carriers, such as cationic lipids and polymers, offer various advantages regarding biosafety and pharmaceutical issues, however they suffer from low efficiency compared to their viral counterparts[69-71,295]. Since efficiency, but also toxicity are depending on the structure and the molecular weight of a polymer and also on the type of the nucleic acid, that has to be delivered, polycations that are supposed to deliver nucleic acids have to be carefully selected for optimal siRNA delivery.

The ability of siRNA to knockdown essentially any gene of interest has become a major focus of interest just in recent times in order to identify important therapeutic genes and develop siRNA based treatments[6-7,12-17]. This is particularly interesting for cancer therapy, where a large number of disease related genes have been explored[296-297]. However, the successful application of siRNA still represents a great challenge. Although many efforts have been already investigated in the field of gene delivery, it is known from literature that efficient carriers of plasmid DNA are not necessarily effective for siRNA delivery[276-277]. Therefore excellent gene silencing activity associated with low toxicity both in vitro and in vivo are the major considerations regarding the design of novel synthetic carrier systems optimized for siRNA delivery.

4.3 Evaluation of modified PEIs with reduced toxicity as efficient siRNA carriers

Polyethylenimine has emerged as one of the most widely used non-viral carrier systems that have been developed for in vitro and in vivo delivery of nucleic acids, as it owns several attributes which are necessary for efficient delivery, such as nucleic acid complexation, protection from nucleases, cell internalization and even endosomal buffering capacity. Nevertheless, high molecular weight PEI, which is a powerful agent for gene transfer[100 103], shows only limited efficiency in siRNA mediated gene silencing[276-277]. Therefore aim of the current study was, in order to improve the properties of the formulation for use in siRNA delivery, to generate a series of branched PEI modifications, which showed far lower toxicity in comparison to unmodified branched PEI and high efficiency in siRNA delivery resulting in powerful knockdown activity[160].

4.3.1 Improved biological properties of modified PEIs

Since toxicity of PEI is mainly associated with the high positive surface charge of polycations, masking of PEI amines was performed in different ways. A relatively slight modification represents the Michael addition of ethyl-acrylate (EA), which results in transformation of the primary amines of PEI into secondary. Moreover, a bulky EA group introduced to the polymer structure could also serve as a steric barrier for interaction of PEI with negatively charged surfaces and, hence, reduce unspecific toxicity of the carrier system. The Ac modification reduces the positive charges of PEI due to acetylation of primary amines. In the Prop series, the transformation of primary into secondary amines was combined with the introduction of negatively charged propionic acid residues to the polymer. Finally, the succinylated (Suc) series represent the strongest reduction of the positive charges of PEI by combination of acylation of primary amines and introduction of negative charges into the polymer.

Importantly, one feature that has to be complied by the conjugates is high siRNA binding affinity and, moreover, the ability to form stable polyplexes. As siRNA binding ability of polymers is strongly influenced by the charge density, this biophysical characterization provides preliminary information about the originated structures. Generally, there exist two methods to examine polyplex formation. Ethidium bromide fluorescence induced by intercalation into double stranded nucleic acids is strongly reduced when polycation/siRNA interaction results in siRNA condensation. Thus, this method can be utilized to monitor siRNA binding affinity in order to obtain information about the density of the polyplexes. Furthermore, complex formation and binding stability can be analyzed by agarose gel electrophoresis. As neutrally or positively charged polyplexes prevent shifting of siRNA in electrical fields, this method is suitable to monitor the complex integrity. In this study, it was shown that modified PEIs still resulted in polycationic structures which effectively bound to siRNA (Figure 5, 6). Nevertheless, gel retardation studies revealed that siRNA binding stability was strongly dependend on the modification degree, as the highest surface modifications showed lowest siRNA complexation stability due to the accelerating loss of protonable nitrogens.

With regard to the toxicity of polymers, as expected, modifications demonstrated strong influence resulting in great reduction of cytotoxicity compared to non-modified polymers (Figure 7). Even the softer PEI modifications, i.e. EA and Ac series, strongly reduced the toxicity, although to a lower extent in comparison to the Prop and Suc series. The observed results are well in accordance with literature, where several modifications of branched PEI resulted in decreased charge density and lower toxicity of the polymer[298-299] due to reduced interactions of polycations with lipid bilayers inducing nanoscale pore formation and thus to a certain extent membrane disruption[114,300].

4 DISCUSSION

The reduction of positive charges, however, might also reduce the activity of polycations in nucleic acid transfer as previously reported[180,271,301]. However, within our series of modifications, only two products with the highest modification degree showed inactivation for siRNA transfer, namely PEI-Prop-52% and PEI-Suc-20% (Figure 8). Reduced polyplex stability due to the increased density of negative charges on the cationic polymer structure is presumably responsible for this, as for example, polyplexes of siRNA and PEI-Suc-20% dissociated in physiological saline (Figure 12). Importantly, all other modified PEIs were found to be able to induce efficient siRNA mediated knockdown of target gene expression, in contrast to unmodified PEI. Among these, PEI-Suc-10% was found to be the most effective delivery agent for siRNA. Low toxicity and high knockdown activity, also at low siRNA concentrations of 50nM, make this PEI modification very promising for in vitro delivery of siRNA.

4.3.2 Structural requirements for efficient siRNA delivery

Many factors, such as charge density or functional groups, strongly influence the delivery process and, thus, the efficiency as well as toxicity of resulting nucleic acid polyplexes. Possible reasons for the low efficiency of unmodified PEI in siRNA delivery, despite its high gene delivery efficiency, might be associated with the stability of polyplexes. Generally, requirements for the delivery of pDNA and siRNA are different. While gene transfer requires nuclear uptake with subsequent unpacking and release of the plasmid, relatively strong binding of the carrier to the pDNA might prevent early unpacking and degradation of pDNA in the cytoplasm. In contrast, for siRNA delivery an effective dissociation from the carrier is necessarily required within the cytoplasm and, thus, a lower binding affinity could be advantageous. Therefore, modifications in order to achieve effective siRNA delivery should tend on the one hand at a lower affinity of the polymer to siRNA in comparison to pDNA polyplexes, on the other hand lower affinity of siRNA to the carrier as compared to pDNA may be also the reason for insufficient knockdown due to polyplex instability[159]. According to the polyplex stability assay, the efficiency of the formulation was found to be independent from the stability of the modified PEI polyplexes with siRNA (Figure 8, 12). Although all PEI modifications in this study were expected to reduce the electrostatic affinity to nucleic acids, certain series, i.e. Ac and Suc, showed even higher stability of polyplexes in comparison to unmodified PEI. Taken into account that all modifications were able to induce significant knockdown, the influence of PEI modifications on the polyplex stability might not be the case for enhanced siRNA delivery efficiency of the formulations. The stability of siRNA formulations may possibly have an important impact with respect to in vivo applications;

however, for in vitro efficiency the extent of polyplex stability has minor importance, unless the polyplex is sufficiently stable to survive the cell culture medium.

The most probable reason for the improved efficiency of modified PEIs in siRNA delivery compared to unmodified PEI is largely a consequence of the lower polymer toxicity. In order to achieve significant knockdown of target genes, PEI based formulations have to be applied at higher concentrations, which are required for sufficient accumulation and "proton sponge" effects in endosomes. This hypothesis is also supported by siRNA delivery studies with unmodified PEI in the presence of additional free less toxic modified polymer (Figure 11), which is separately internalized into endocytic vesicles[278]. Higher amounts of PEI finally in the endosomal compartments may increase escape to the cytoplasm due to a stronger "proton sponge" effect[105-108], before the fusion of endocytotic vesicles with primary lysosomes takes place exposing the polyplexes to degradation by lysosomal enzymes. Such polymer amounts could not be applied for unmodified PEI without high carrier toxicity, whereas the far less toxic modified analogues could be applied in significantly higher concentrations facilitating efficient knockdown of target genes without any side effects. At these concentrations, even small amounts of siRNA were able to decrease the activity of the target gene expression. Therefore, the reduction of the toxicity has a really decisive effect in vitro on the efficiency of siRNA formulations with modified PEIs.

4.4 Evaluation of biodegradable OEI conjugates for siRNA delivery

Currently, biodegradable polymers have emerged as most promising candidates for the development of non-viral carrier systems with improved efficiency in nucleic acid delivery and better biocompatibility[45,111,119-136]. Reduced toxicity is undoubtedly one of the most important aspects with regard to in vitro and in vivo applications of synthetic nucleic acid carrier systems, while at the same time high delivery efficiency has to be achieved. Generally, toxic properties increase with the number of positive charges, the hydrophobicity and the molecular weight of the carrier systems[109]. Neutralization of the positive polymer charges by chemical modification is one way to obtain reduced acute toxicity, both on the cellular level and in the organism, however does not eliminate the long-term fate of the polymeric carrier in the host, which has to be also taken into consideration[302]. Non-degradable high molecular weight polymers, such as PEI, are more or less static structures which often cannot be metabolized and tend to accumulate mostly in the liver or kidney[303] resulting in undesired long-term toxicity. Therefore, the main focus was to generate a variety of novel biocompatible polymers of adequate sizes, which allow effective siRNA delivery and are well tolerated by the host organism due to their degradability into smaller non-toxic decomposition products. Notably, degradability in general and degradation kinetics will rather impact the long-term toxic effects of the carrier systems than having an explicit influence on the acute cytotoxicity.

4.4.1 Oligomerized OEIs: targeting for optimized virus-like siRNA delivery

Basically, non-degradable polycations with increased molecular weight provide strong polyplex stability and exhibit in most cases enhanced delivery efficiency of nucleic acids[304], but also higher toxicity than their low molecular weight counterparts, as they are hardly eliminated by the organism[302]. Based on these findings, the advantage of low molecular weight polymers, i.e. low toxicity, was combined with the advantage of high molecular weight polymers, i.e. high efficiency, by bioreversibe crosslinking of low molecular weight polymers. In a combinatorial approach, oligomerization of small polycations via bifunctional linkers into larger molecules was devised in a way of increasing molecular weight and, thus, nucleic acid binding capacity while preserving their beneficial non-toxic nature. Several approaches have been already described in literature for gene delivery dealing with crosslinking of low molecular weight polyamines via various degradable linkages in order to reduce toxicity and maintain high efficiency in vitro and in vivo[111,119-120,123-125,127-134].

Thus, an important purpose of this study was to explore novel biodegradable polycations, which are highly effective in siRNA delivery and could be easily metabolized and eliminated by the host resulting in less toxic side effects. Thereby promising results have been obtained by propionamide crosslinked low molecular weight oligoethylenimines, which were able to bind effectively to siRNA (Figure 14) providing the beneficial low long-term toxic properties of OEI 800 and high efficiency for siRNA delivery (Figure 15). The degradation pattern of the originated amid bond containing OEI-HD-1 polymers, which are potentially enzymatically degradable, was confirmed based on time-dependent amide hydrolysis[274].

In cancer therapy high tumor specificity can be achieved in several ways, ranging from direct delivery of polyplexes into the target site, physical strategies, such as electroporation[305-307], magnetofection[308-311], sonoporation[312-314] or photodynamic therapy[129,315-319], up to utilization of targeting ligands for specific uptake into the cells of interest. In this study, in order to convert OEI-HD-1 polyplexes into even more biocompatible particles with high tumor specificity, transferrin was selected as cell surface targeting ligand, since transferrin receptors are overexpressed in a variety of tumor cells due to their higher demand of iron, needed for the fast growth[192]. Similar strategies have been already successfully applied by us and other researchers for specific tumor targeting of DNA[116,173] and siRNA[193] complexes. Moreover, transferrin as protein ligand combines both a targeting function towards the tumor cells and a shielding function of the polyplexes reducing unspecific interactions with non-target sites[191,194].

Thus, the incorporation of Tf as a targeting ligand into OEI-HD-1 polyplexes in order to enhance the delivering specificity might a further step towards the design of more virus-like and biodegradable synthetic carrier systems. Investigations of the optimized Tf containing formulations in vitro displayed efficient knockdown activity in Neuro2A/eGFPLuc cells (Figure 19), which are known to overexpress the Tf receptor on the cell surface[194,199]. Specific uptake via the Tf receptor was confirmed by competitive inhibition of the Tf receptor with an excess of free Tf ligand. Blocking the Tf receptor resulted in reduced reporter gene silencing activity of Tf targeted OEI-HD-1 polyplexes, while the knockdown mediated by siRNA formulations without targeting ligand remained unaffected due to unspecific uptake of these formulations.

Notably, in contrast to the standard positively charged polyplexes, the transferrin modified formulations showed reduced zeta potentials preventing aggregation at lower w/w ratios, which makes them even more suitable for in vivo applications[40]. Further evaluation of the therapeutic potential of OEI-HD-1 formulated siRNA in vivo was performed by Nicole Tietze as part of her PhD thesis (LMU, 2009).

4.4.2 Pseudodendritic oligoamines with high potential for siRNA delivery

In a further approach, another class of biodegradable synthetic nucleic acid delivery systems, namely pseudodendrimers, was evaluated for their siRNA delivery potential, which also feature less toxic side effects compared to standard non-degradable high molecular weight polyethylenimine. Basically, the efforts to generate biocompatible polymers based on crosslinked low molecular weight polycations, as described previously, exhibited the disadvantage of a relatively high polydispersity regarding the resulting compounds, which implies that they maintain a stronger heterogeneity in their molecular weight distribution caused by the polymer synthesis procedure. Thus, an improvement towards better defined polymers was achieved in the design of pseudodendritic structures by the use of branched low molecular weight oligoethylenimine as core building block resulting in closer molecular weight distribution of the arising conjugates. Different dioldiacrylates containing hydrolysable ester bonds, were applied to form dendritic branches, which were additionally modified with various oligoamines in order to generate versatile pseudodendritic structures. The obtained conjugates exhibited molecular weights of about 4 - 8kDa containing high amounts of hydrolysable ester bonds[135].

In vitro studies showed that siRNA delivery efficiency and cytotoxicity were dependent on both pseudodendritic core characteristics and surface modification. Thus, enhancing the

4 DISCUSSION

pseudodendritic core hydrophobicity from OEI-ED to OEI-HD core conjugates resulted in increased cytotoxicity (Figure 24). Furthermore, the different surface amines of OEI-HD cores, but not in case of OEI-ED or OEI-BD cores, also showed an influence on cytotoxicity, as OEI surface modified OEI-HD core conjugates exhibited slightly decreased toxicity of polyplexes.

Regarding reporter gene silencing activity, only Sp and S surface modifications within the OEI-HD core conjugates were able to mediate efficient knockdown of luciferase expression without unspecific toxicity. These results indicate that both the pseudodendritic core characteristics along with the surface modification within polyplexes has an important impact on high siRNA delivery activity and cytotoxicity as demonstrated in the OEI-HD core conjugates. Obviously, both cytotoxicity and siRNA delivery efficiency depend on an optimized balance between hydrophobic and hydrophilic domains within the pseudodendrimers, as for example merely hydrophilic structures like OEI-ED core conjugates resulted in low cytotoxicity but also none knockdown efficiency.

The reason for why only Sp and S modified OEI-HD core conjugates showed remarkable knockdown efficiency could be an intrinsic endosomolytic property of these conjugates as reported by Julia Klöckner (PhD thesis 2005, LMU), which is highly desirable for efficient release of the polyplexes from intracellular vesicles. One possible explanation for this membrane lytic activity may be their proposed amphiphilic micelle-like structure resulting from a hydrophilic core, which is assembled with various hydrophobic HD residues that are again linked to hydrophilic spermidine or spermine moieties. This design might promote interactions with lipid bilayers and, thus, destabilization and disruption of cellular membranes. During the extracellular delivery process, high lytic activity is strongly undesired and considered as toxic side effects, on the other hand within the cell, membrane destabilization characteristics are beneficial for efficient endosomal release of the polyplexes[243].

Taken into account that polymer concentrations responsible for cellular destruction are expected to be higher, than needed for endosomolytic activity, for in vivo applications there would be a therapeutical range of polymer doses, which could potentially trigger endosomal release without unspecific toxicity.

4.4.3 Hydrophobically modified OEIs: structural influence on biological activity

The development of polymeric carrier systems for siRNA delivery is mainly attributed to high molecular weight polyamines, which often represent a problem for in vivo application due to their relatively narrow therapeutic window. Being either non-degradable as in case of PEI 25, or showing relatively slow degradation kinetics, such structures may lead to accumulation of toxicity after repeated administration. In contrast, short oligoamines are far less toxic[109,320] and can be rapidly excreted from the body. As such formulations, however, show insufficient stability in blood circulation and low siRNA delivery efficiency, mainly due to poor endosomolytic properties, an alternative approach was evaluated. Instead of crosslinking into high molecular weight structures, low molecular weight oligoamines were modified with hydrophobic moieties[290,321-322]. Such a concept has several advantages: (I) these formulations are supposed to have low half-life times in the organism, presumably due to their low molecular weight, which is favorable for excretion and metabolism, (II) the hydrophobic interactions stabilize polyplexes during storage and administration, (III) similarly to endosomolytic peptides, certain modifications enhance interactions with lipid bilayers in order to promote transfer across cellular membranes of intracellular vesicles, e.g. by lysis of endosomes.

For that purpose, low molecular weight oligoethylenimine 800Da was modified with different alkyl-acrylates[275], which offer the advantage of relatively rapid enzymatic degradation of ester bonds in the body and renal clearance of the metabolites. Even in physiological pH buffer without enzymes, the degradation of ester bond proceeds relative quickly (Figure 27, 31).

Moreover, the incorporation of longer alkyl chains with higher hydrophobicity, i.e. HA and LA modification, generally increased the stability of polyplexes in salt containing buffers due to stronger hydrophobic interactions among each other. On the other hand the incorporation of shorter alkyl chains with lower hydrophobicity, i.e. EA and BA modification, led to a decrease of binding affinity to siRNA, most probably due to steric barriers between the charges of siRNA and OEI introduced by a bulky alkyl group (Figure 28). The siRNA formulations with EA as well as BA modified OEIs were also quite unstable against aggregation even in salt-free HBG buffer (Table 6).

Gene silencing experiments revealed that OEI-HA-10 was the only effective oligoamine for siRNA delivery (Figure 32). All other potentially active formulations either precipitated during preparation or were not able to be internalized due to the large size of the polyplex particles.

Among all oligoamines, only BA and HA modifications could effectively induce erythrocyte lysis (Figure 35). Surprisingly, oligoamines with long hydrophobic chains, i.e. LA modification, did not show any ability to lyse cellular membranes. Even OEI-LA-5, despite similar hydrophilic/hydrophobic balance as OEI-HA-10, was rather ineffective in erythrocyte lysis and consequently in siRNA delivery.

However, due to the strong lytic activity, OEI-HA-10 formulations were relatively toxic in vitro, which led to enhanced acute toxicity and lethality during in vivo applications in mice performed by Nicole Tietze as part of her PhD thesis (LMU, 2009).

Therefore, a promising optimization procedure was a dilution of OEI-HA-10 in the formulation by far less toxic but ineffective oligoamines, such as OEI 800 and OEI-LA-5. Gradual decrease of the OEI-HA-10 content resulted in strongly decreased cytotoxicity of the formulation, while importantly no decrease of the efficiency was observed. The co-formulation with OEI-LA-5 showed additional advantage of colloidal stabilization, which could be potentially important in view of in vivo applications, while incorporation of OEI 800 to the polyplexes decreased both colloid stability and stability against dissociation (Figure 29, Table 7).

The co-formulations with different lipophilic agents, such as DOPE, DOPC and DPPC, were also able to decrease the toxicity of the formulations in vitro. However, in most cases, except DOPE, these led to inactivation of the formulations, as the strongest decrease of toxicity corresponded to the strongest inactivation of the formulations. The OEI-HA-10/DOPE formulations showed only slightly less toxicity in comparison to OEI-HA-10 alone, which could be involved with certain restrictions for in vivo application of such formulations, but notably the siRNA delivery efficiency, however, was greatly improved in vitro. Due to the inverse head-tail symmetry of DOPE lipid, it is known to destabilize the lipid bilayer structure after incorporation into lipid membranes. Thus, the endosomolytic character of the DOPE lipid seems to be the main reason for the improved efficiency of the OEI-HA-10/DOPE formulations.

Such optimized formulations with greatly improved biocompatibility and efficiency, if stable in the bloodstream, could potentially represent a promising siRNA delivery approach for in vivo applications that warrants further investigation.

4.5 Evaluation of bioresponsive endosomolytic conjugates for siRNA delivery

A key issue in the field of nucleic acid delivery remains the development of dynamic and bioresponsive polymers, so called "artificial viruses"[70,169-171]. Therefore, the sophisticated mechanisms that viruses have developed[323] to overcome the barriers they are confronted with inside their host[152] can serve as a guide towards the design of more flexible and stimuli responsive nucleic acid delivery systems.

Especially toxicity and poor endosomal release limit the application of non-viral carrier systems, whereby a variety of endosomolytic peptides has been evaluated for the improvement of nucleic acid delivery[225-226]. In this study, the cationic membrane-active peptide melittin was utilized, displaying high lytic activity on cellular membranes, which is, however, unfavorable in the extracellular environment as it mediates toxic side effects[240]. Hence, dimethylmaleic anhydride was used to mask the lytic activity of melittin at neutral pH which can be restored again after acidic cleavage of the protecting groups[265]. Maleic anhydrides have the property to reversibly react with the lysine residues and the N-terminal amino group of peptides and to be removed again after slightly acidification like in endosomal compartments[324-325].

4.5.1 DMMAn-Mel modification for enhanced siRNA delivery efficiency

In order to evaluate the effect of pH specific membrane lytic activity on siRNA delivery efficiency, DMMAn masked melittin peptides were covalently attached to siRNA binding polycations[247]. For that purpose, PLL, PEI and OEI-HD-1 were chosen as backbone polymers. Biodegradable PLL shows no intrinsic endosomal escape mechanism, which enables an isolated consideration of the endosomolytic effect of melittin. In contrast, the non-degradable PEI as one of the most frequently used polycations for nucleic acid delivery allows a combined effect consisting of the lytic activity of melittin and the intrinsic endosomal escape properties of PEI due to its "proton sponge" capacity. Finally, oligoethylenimine based OEI-HD-1 represents a biodegradable polymer like PLL, possesses endosomal escape properties like PEI and, moreover, shows inherent siRNA delivery efficiency, which might result in strong synergistic effects with the membrane destabilizing activity of melittin.

In order to improve solubility and avoid undesired polyplex aggregation, PEG with an average molecular weight of 5kDa was primarily grafted onto the polycation prior to peptide coupling. Although it is known that PEGylation can almost neutralize the surface charge of polyplexes[245], PEG with a higher molecular weight, e.g. 20kDa, has to be applied for such purpose. The remaining positive surface charge of the polyplexes strongly contributes to the interaction with cellular membranes, which is important for internalization, as none additional targeting ligand is included within the formulation.

Gene silencing experiments revealed that all DMMAn-Mel modified polyplexes showed remarkable improved knockdown efficiency compared to the other conjugates (Figure 40, 42, 44). While PLL and PEI are ineffective in siRNA delivery, OEI-HD-1 shows proper knockdown activity, which was greatly enhanced after coupling with the endosomolytic melittin peptides. PEGylation alone did not alter siRNA delivery activity of all formulations.

Apart from poor endosomal release also cytotoxicity of carrier systems is one of the limiting barriers for siRNA delivery. In particular, high lytic activity of the polymers before reaching their target site of action raises this problem. Notably, DMMAn-Mel modified conjugates were much better tolerated by cells resulting in higher metabolic activity than their unmodified or PEGylated counterparts (Figure 40, 42, 44). These data indicate that DMMAn grafting of melittin not only avoids the undesired lytic activity in the extracellular environment, but also further reduces acute toxicity of the polycationic carriers. Presumably the ability of

4 DISCUSSION

polycations to interact with cellular membranes resulting in nanoscale pore formation is strongly decreased due to the additional negative charges of the DMMAn groups, which mask the regular polycationic positive charges and, hence, minimize interactions with lipid bilayers[114,300].

Monitoring the hemolytic activity of DMMAn-Mel modified conjugates in an erythrocyte leakage assay resulted in reduced lytic activity at physiological pH. However, the lytic activity could be entirely restored after preincubation at endosomal pH due to the cleavage of the DMMAn groups (Figure 45). These results demonstrate that DMMAn protecting groups are able to reversible mask the lytic activity of melittin, and, thus, can be utilized to generate bioresponsive conjugates with the desired lytic activity profile.

In summary PEGylation and DMMAn-Mel modification of polycations allowed the formation of nanosized polyplexes with strongly enhanced siRNA delivery efficiency compared to their unmodified counterparts.

4.5.2 Dynamic siRNA-polymer conjugates for programmed delivery

In order to survive the extracellular delivery process, siRNA has to remain stable associated with the carrier system until they reach the site of action within the target cell. Basically, most lipoplex and polyplex formulations are kept together via electrostatic interactions. The risk therefore lies in the disruption of such complexes due to undesired interactions with other charged molecules before they reach the target cells. It was already shown in literature, that both the serum and the extracellular matrix can lead to carrier disassembly[157-158], which consequently negatively affects the siRNA delivery efficiency, especially in view of in vivo applications. Carrier unpacking is rather a problem concerning siRNA than pDNA, due to the lower number of negative charges per nucleic acid molecule, i.e. lower stability of siRNA polyplexes[159-160].

An elegant approach to compensate this deficiency was achieved by covalent attachment of siRNA to the cationic polymeric carrier via a reversible disulfide linkage, which enables dissociation of polyplexes and release of the nucleic acid within the reductive intra-cellular environment[85]. Luciferase siRNA and control siRNA were each coupled to PEG-PLL-DMMAn-Mel conjugates containing all the functional domains in one molecule, i.e. PEG for improved solubility, PLL as polycation, pH responsive lytic peptide melittin and bioreversible attached siRNA. Gene silencing experiments with electrostatic assembled siRNAs already revealed that the polycation to siRNA ratio has a great impact on delivery efficiency and cytotoxicity. In order to compare the covalently attached siRNA-polymer conjugate with the electrostatic analogue, a PLL to siRNA ratio of about 2/1 (w/w) was chosen, which was found to mediate strong gene silencing activity in case of the electrostatic PEG-PLL-DMMAn-

4 DISCUSSION

Mel/siRNA polyplexes. This ratio results in an overall net positive charge of the conjugate, which is on the one hand necessary for cell interaction and consequently internalization, as none additional targeting ligand is included within the formulation, and on the other hand useful for additional complexation of the siRNA in order to prevent enzymatic degradation, e.g. by serum RNases.

To evaluate the gene silencing capability of covalently attached PEG-PLL-DMMAn-Mel-siRNA conjugates, siRNA delivery activity was directly compared with the electrostatic assembled PEG-PLL-DMMAn-Mel/siRNA polyplexes formed at w/w ratio of 2/1 (Figure 48). High comparable gene knockdown activity was observed for both conjugates in a dose dependant reduction of luciferase expression, even at lowest siRNA doses, indicating that covalent attachment of siRNA did not weaken the siRNA delivery efficiency. Conjugates containing electrostatically or covalently attached control siRNA did not negatively affect the luciferase expression, indicating specific siRNA mediated gene silencing.

Additionally, significant carrier toxicity was very low for both conjugates, as reduced metabolic activity of cells began to occur only at the highest siRNA dose, confirming the transfection results obtained with formulations containing control siRNA (Figure 48).

The expected pH responsiveness of the PEG-PLL-DMMAn-Mel-siRNA conjugate was proven in an erythrocyte leakage assay (Figure 50). Consistent with the acidic cleavage of the DMMAn masking groups from melittin, incubation of the conjugate at endosomal pH strongly enhanced the lytic activity.

To further investigate the intracellular delivery process of the siRNA-polymer conjugate, the ability of glutathione to induce cleavage of the disulfide bond facilitating siRNA release from the carrier system was analyzed via agarose gel electrophoresis. The natural reducing agent glutathione, a tripeptide consisting of glutamate, cysteine and glycine, is excessive present in the intracellular compartment. Thus, to clarify if physiological cytoplasmatic glutathione concentrations can cleave the siRNA connecting disulfide bonds, conjugates were incubated for 1h at 37°C with appropriate glutathione concentrations. Additionally, the natural polyanion heparin, a highly sulfated glycosaminoglycan closely related in structure to heparin sulfates in the extracellular matrix, was applied in order to dissasemble non-covalent siRNA polyplexes ensuring that free siRNA is released from the conjugate[85]. However, without reductive intracellular environment and cleavage of disulfide bonds, also heparane sulfates as well as other negatively charged macromolecules cannot lead to carrier disassembly. As shown, physiological glutathione concentrations (1.25 - 5mM) in combination with heparin were able to release the siRNA from the conjugate (Figure 51). Based on these observations, certain cleavage of the siRNA before formation of the RNA-inducing silencing complex is expected.

4 DISCUSSION

However, it still remains unclear, if the cleavage of the disulfide linkage is required for biological activity of the PEG-PLL-DMMAn-Mel-siRNA conjugates. Literature indicates a possible advantage of bioreducibility, as for example quantum dots containing covalently attached siRNA, showed greater silencing efficiency when siRNA was reversibly attached by disulfide linkers in comparison to non-reducible thioether linkers, suggesting that disulfide cleavage in the intracellular environment takes place and release of siRNA is beneficial[83,326]. Similar observations have been also described where it was also found out that a bioreversible siRNA linkage is advantageous in terms of gene silencing activity[84].

In summary, the bioresponsive endosomolytic PEG-PLL-DMMAn-Mel-siRNA conjugates belong to a new generation of dynamic, multifunctional nucleic acid carrier systems, which are able to undergo molecular changes triggered by the physiological environment[242-244,264,269,327]: (I) covalent attachment of siRNA improves the stability of the siRNA formulations in the extracellular environment, (II) the endosomal pH is expected to recover the lytic activity of melittin required for efficient release from the endosome, (III) in addition to the dynamic endosomolytic activity, the reducing intracellular environment triggers programmed cleavage of the disulfide linkage and enables release of the siRNA within the cytoplasm. Consistently, PEG-PLL-DMMAn-Mel-siRNA conjugates showed excellent in vitro gene silencing activity comparable to the electrostatically formed siRNA polyplexes.

However, despite the very encouraging bioactivity and biocompatibility in vitro, intravenous and intratumoral in vivo applications in mice showed unexpectedly high acute toxicity with the current formulation. The reasons for this are currently not completely clear, but lack of targeting functionality, incomplete PEG shielding and conjugate aggregation may contribute to the pronounced in vivo toxicity. Thus, in order to exploit the full potential of the concept, optimized siRNA-polymer conjugates with targeting ligands and improved shielding moieties remain to be generated for further evaluation.

5 SUMMARY

Small interfering RNA molecules offer a promising tool for the treatment of various diseases including cancer and a number of other inherent or acquired disorders, due to their ability to knockdown essentially any target gene of interest. However, the successful application of siRNA based therapy still represents a great challenge. As optimization of the delivery strategy remains one of the major restrictions for clinical use, research has progressively focused on the development of synthetic delivery devices suitable for safe and specific delivery of siRNA in order to exploit the huge potential of RNA interference. This thesis reports the discovery and optimization of novel polymers as highly effective and biocompatible siRNA delivery systems.

Exploring the strong potential of polyethylenimine, a powerful gene delivery agent but far less effective in case of siRNA, various optimized less toxic PEI derivates were screened for the delivery of siRNA. Studies on their biophysical and biological characteristics revealed that modifications which reduce the highly positive surface charge of the polymer resulted in strongly decreased toxicity and a better therapeutic window of the formulation. Gene silencing studies demonstrated remarkable knockdown of target gene expression, even when using only small amounts of siRNA. It was elucidated that the stability of siRNA polyplexes had only marginal importance for knockdown activity in vitro. In contrast, reduced polymer toxicity exhibited crucial impact on the efficiency of siRNA delivery due to the applicability of higher concentrations of PEI based formulations, which are required for sufficient accumulation and "proton sponge" effects in endosomes. Thus, efficient release of the polyplexes into the cytoplasm is provided and results in high siRNA activity avoiding the undesired degradation by lysosomal enzymes.

Application of non-degradable carrier systems, however, often leads to accumulation of toxicity in vivo, which narrows the therapeutic window and the success of nucleic acid based therapy. Thus, biodegradation of the carriers is a desired property which has to be taken into consideration in the design of novel polymers for effective treatment in vivo. Covalent crosslinking of low molecular weight polycations with biodegradable linkers into high molecular weight polycations is one strategy to achieve nucleic acid carrier systems capable of beeing degraded into smaller fragments in the appropriate cellular microenvironment. In this approach, promising results were found for propionamide crosslinked low molecular weight oligoethylenimines, which were highly effective in siRNA delivery. Furthermore, in order to optimize these conjugates into virus-like siRNA delivery systems, transferrin as site-specific targeting ligand was successfully incorporated into the carrier systems providing the opportunity of increased target-specific delivery, which is primarily essential for cancer

5 SUMMARY

therapy in vivo. The concept was further developed by generating polymers exhibiting a better defined chemical structure, so-called pseudodendrimers. These conjugates were formed by adding several monomeric moieties upon a central core unit via biodegradable linkages in order to obtain pseudodendritic structures. In vitro structure-activity relationship studies revealed that siRNA delivery efficiency and cytotoxicity were dependent on both pseudodendritic core characteristics and surface modification. Only spermine and spermidine surface modifications within the pseudodendritic cores with highest hydrophobicity emerged as highly effective siRNA delivery agents, presumably due to an optimized balance between hydrophobic and hydrophilic domains within the pseudodendrimers, which is required for highest siRNA delivery efficiency at low cytotoxicity. In an alternative approach, instead of crosslinking, low molecular weight polycations were modified with hydrophobic moieties. In vitro studies demonstrated that, in particular, oligoethylenimine modified by ten hexyl acrylates was the most promising siRNA delivery system facilitating increased stability of polyplexes against dissociation and improved colloidal stability due to hydrophobic interactions between the OEI chains. The lytic properties of these conjugates ensured effective intracellular transport across cellular membranes, presumably by promoting sufficient escape from endosomes. In addition, different co-formulation strategies, including helper polymer-based and lipid-based modifications, could greatly improve the biocompatibility and efficiency of the formulation.

Apparently, poor endosomal escape represents a major barrier for successful siRNA delivery. For this reason, in order to avoid degradation in the acidic endosomal compartment, different polymeric carriers based on PEI, PLL and oligomerized OEI, respectively, were conjugated with the membrane active peptide melittin to escape endosomal entrapment. With the intention to prevent undesired membrane destabilizing properties in the extracellular environment, the lytic activity of melittin was reversibly masked with a pH responsive protecting group. These carriers act more dynamically in response to their cellular microenvironment due to a triggered lytic activity of melittin only upon acidifaction in the endosome mediating greatly enhanced efficiency of the formulations for siRNA delivery in vitro, while toxicity was remarkable reduced. However, the trend of these formulations to aggregate required modification with polyethylene glycol for improved solubility and stabilization. Principally one weakness of electrostatically associated polyplexes is due to the fact that other physiological biomolecules can disrupt such complexes, which results in vector disassembly before reaching the target site. To ensure stable association during the extracellular delivery process, siRNA was covalently attached to the carrier system by bioreducible disulfide linkers capable of facilitating release of siRNA in the cytoplasm, which is of crucial importance for RISC assembly and, thus, gene silencing activity. Taken together, such pre-programmed bioresponsive systems promoting intracellular release of siRNA in the

5 SUMMARY

cytoplasm represent a considerably step in the optimization process. Combination with targeting ligands and distinct shielding elements will be an additional emphasis for the development of efficient delivery devices for cancer therapy.

6 ABBREVIATIONS

Ac	acetic anhydride
asRNA	antisense RNA
BA	butyl-acrylate
BD	butanediol-diacrylate
cDNA	complementary DNA
CMV	cytomegalovirus
DLS	dynamic light scattering
DMEM	Dulbecco's Modified Eagle's Medium
DMMAn	dimethylmaleic anhydride
DMSO	dimethyl sulfoxide
DNA	deoxyribonucleic acid
DOPC	dioleoylphosphatidylcholine
DOPE	dioleoyl-phosphatidylethanolamine
DPPC	dipalmitoylphosphatidylcholine
DTT	dithiothreitol
E	ethanolamine
EA	ethyl-acrylate
ED	ethyleneglycol-diacrylate
EDTA	ethylenediamine tetraacetic acid
eGFP	enhanced green fluorescent protein
EPR	enhanced permeability and retention
EtBr	ethidium bromide
FCS	fetal calf serum
FCS	fluorescence correlation spectroscopy
GSH	glutathione
HA	hexyl-acrylate
HBG	HEPES-buffered glucose
HD	hexanediol-diacrylate
HEPES	N-(2-hydroxyethyl) piperazine-N'-(2-ethansulfonic acid)
HPLC	high pressure liquid chromatography

6 ABBREVIATIONS

LA	lauryl-acrylate
Luc	luciferase
Mel	all-(D)-melittin (with a cysteine residue at the N-terminus)
miRNA	micro RNA
mPEG	monomethoxy PEG
mRNA	messenger RNA
MTT	dimethylthiazolyldiphenyl-tetrazolium bromide
MW	molecular weight
^{1}H-NMR	nuclear magnetic resonance
N/P-ratio	molar ratio of nitrogen to phosphate (conjugate to nucleic acid) (PLL: molar ratio of epsilonamino nitrogen to phosphate)
OEI	oligoethylenimine
OEI 800	OEI with an average molecular weight of 800Da
PBS	phosphate-buffered saline
pDNA	plasmid DNA
PDP	pyridyldithio propionate
PEI	polyethylenimine
PEI 22	linear PEI with an average molecular weight of 22kDa
PEI 25	branched PEI with an average molecular weight of 25kDa
PEG	polyethylene glycol
PLL	poly-L-lysine with an average molecular weight of 32kDa
PLL50	PLL with 50 lysine monomer units
PLL185	PLL with 185 lysine monomer units
polyIC	poly-inosine-cytosine
Prop	propionic acid
RISC	RNA-induced silencing complex
RLU	relative light units
RNA	ribonucleic acid
RT-qPCR	reverse transcriptase quantitative real time quantitative polymerase chain reaction
S	spermine
shRNA	short hairpin RNA
SEC	size exclusion chromatography

6 ABBREVIATIONS

siRNA	small interfering RNA
Sp	spermidine
SPA	succinimidyl propionic acid
SPDP	succinimidyl 3-(2-pyridyldithio) propionate
Suc	succinic anhydride
TBE	tris borate EDTA
Tf	transferrin
TNBS	trinitrobenzenesulfonic acid
w/w ratio	weight to weight ratio (conjugate to nucleic acid)

7 REFERENCES

1. Kircheis, E. Ostermann, M.F. Wolschek, C. Lichtenberger, C. Magin-Lachmann, L. Wightman, M. Kursa and E. Wagner: Tumor-targeted gene delivery of tumor necrosis factor-alpha induces tumor necrosis and tumor regression without systemic toxicity, Cancer Gene Ther 2002; 9(8):673-680.
2. A.T. Stopeck, A. Jones, E.M. Hersh, J.A. Thompson, D.M. Finucane, J.C. Gutheil and R. Gonzalez: Phase II study of direct intralesional gene transfer of allovectin-7, an HLA-B7/beta2-microglobulin DNA-liposome complex, in patients with metastatic melanoma, Clin.Cancer Res. 2001; 7(8):2285-2291.
3. M. Bergen, R. Chen and R. Gonzalez: Efficacy and safety of HLA-B7/beta-2 microglobulin plasmid DNA/lipid complex (Allovectin-7) in patients with metastatic melanoma, Expert.Opin.Biol.Ther. 2003; 3(2):377-384.
4. M.L. Edelstein, M.R. Abedi and J. Wixon: Gene therapy clinical trials worldwide to 2007- an update, J.Gene Med. 2007; 9(10):833-842.
5. E. Wagner, R. Kircheis and G.F. Walker: Targeted nucleic acid delivery into tumors: new avenues for cancer therapy, Biomed.Pharmacother. 2004; 58(3):152-161.
6. M.A. Behlke: Progress towards in vivo use of siRNAs, Mol.Ther. 2006; 13(4):644-670.
7. M. Meyer and E. Wagner: Recent developments in the application of plasmid DNA-based vectors and small interfering RNA therapeutics for cancer, Hum.Gene Ther 2006; 17(11):1062-1076.
8. H. Yin, Q. Lu and M. Wood: Effective exon skipping and restoration of dystrophin expression by peptide nucleic acid antisense oligonucleotides in mdx mice, Mol Ther 2008; 16(1):38-45.
9. J. Krutzfeldt, N. Rajewsky, R. Braich, K.G. Rajeev, T. Tuschl, M. Manoharan and M. Stoffel: Silencing of microRNAs in vivo with 'antagomirs', Nature 2005; 438(7068):685-689.
10. M.M. Fabani and M.J. Gait: miR-122 targeting with LNA/2'-O-methyl oligonucleotide mixmers, peptide nucleic acids (PNA), and PNA-peptide conjugates, RNA 2008; 14(2):336-346.
11. A. Fire, S. Xu, M.K. Montgomery, S.A. Kostas, S.E. Driver and C.C. Mello: Potent and specific genetic interference by double-stranded RNA in Caenorhabditis elegans 2, Nature 1998; 391(6669):806-811.
12. G.J. Hannon and J.J. Rossi: Unlocking the potential of the human genome with RNA interference, Nature 2004; 431(7006):371-378.
13. M. Sioud: Therapeutic siRNAs, Trends Pharmacol.Sci 2004; 25(1):22-28.
14. A. Aravin and T. Tuschl: Identification and characterization of small RNAs involved in RNA silencing, FEBS Lett. 2005; 579(26):5830-5840.
15. A.W. Tong, Y.A. Zhang and J. Nemunaitis: Small interfering RNA for experimental cancer therapy, Curr.Opin.Mol.Ther. 2005; 7(2):114-124.
16. A. Aigner: Gene silencing through RNA interference (RNAi) in vivo: strategies based on the direct application of siRNAs, J.Biotechnol. 2006; 124(1):12-25.
17. S.I. Pai, Y.Y. Lin, B. Macaes, A. Meneshian, C.F. Hung and T.C. Wu: Prospects of RNA interference therapy for cancer, Gene Ther. 2006; 13(6):464-477.

7 REFERENCES

18. M. Erdmann, J. Dorrie, N. Schaft, E. Strasser, M. Hendelmeier, E. Kampgen, G. Schuler and B. Schuler-Thurner: Effective clinical-scale production of dendritic cell vaccines by monocyte elutriation directly in medium, subsequent culture in bags and final antigen loading using peptides or RNA transfection, J Immunother 2007; 30(6):663-674.

19. V.F. Van Tendeloo, P. Ponsaerts and Z.N. Berneman: mRNA-based gene transfer as a tool for gene and cell therapy, Curr Opin Mol Ther 2007; 9(5):423-431.

20. S.T. Crooke: Antisense strategies, Curr Mol Med 2004; 4(5):465-487.

21. S.T. Crooke: Progress in antisense technology, Annu Rev Med 2004; 5561-95.

22. P.D. Zamore, T. Tuschl, P.A. Sharp and D.P. Bartel: RNAi: double-stranded RNA directs the ATP-dependent cleavage of mRNA at 21 to 23 nucleotide intervals, Cell 2000; 101(1):25-33.

23. S.M. Elbashir, J. Harborth, W. Lendeckel, A. Yalcin, K. Weber and T. Tuschl: Duplexes of 21-nucleotide RNAs mediate RNA interference in cultured mammalian cells 1, Nature 2001; 411(6836):494-498.

24. A. Shir, M. Ogris, E. Wagner and A. Levitzki: EGF receptor-targeted synthetic double-stranded RNA eliminates glioblastoma, breast cancer, and adenocarcinoma tumors in mice, PLoS Med 2006; 3(1):e6.

25. N. Tomita, H. Azuma, Y. Kaneda, T. Ogihara and R. Morishita: Application of decoy oligodeoxynucleotides-based approach to renal diseases, Curr.Drug Targets. 2004; 5(8):717-733.

26. H. Ulrich, C.A. Trujillo, A.A. Nery, J.M. Alves, P. Majumder, R.R. Resende and A.H. Martins: DNA and RNA aptamers: from tools for basic research towards therapeutic applications, Comb.Chem High Throughput.Screen. 2006; 9(8):619-632.

27. S.M. Elbashir, W. Lendeckel and T. Tuschl: RNA interference is mediated by 21- and 22-nucleotide RNAs 2, Genes Dev. 2001; 15(2):188-200.

28. S.M. Hammond, S. Boettcher, A.A. Caudy, R. Kobayashi and G.J. Hannon: Argonaute2, a link between genetic and biochemical analyses of RNAi, Science 2001; 293(5532):1146-1150.

29. T.A. Rand, K. Ginalski, N.V. Grishin and X. Wang: Biochemical identification of Argonaute 2 as the sole protein required for RNA-induced silencing complex activity, Proc Natl Acad Sci U S A 2004; 101(40):14385-14389.

30. Y. Wang, G. Sheng, S. Juranek, T. Tuschl and D.J. Patel: Structure of the guide-strand-containing argonaute silencing complex, Nature 2008; 456(7219):209-213.

31. D. Bumcrot, M. Manoharan, V. Koteliansky and D.W. Sah: RNAi therapeutics: a potential new class of pharmaceutical drugs, Nat.Chem Biol 2006; 2(12):711-719.

32. L. Aagaard and J.J. Rossi: RNAi therapeutics: principles, prospects and challenges, Adv Drug Deliv.Rev. 2007; 59(2-3):75-86.

33. T. Nguyen, E.M. Menocal, J. Harborth and J.H. Fruehauf: RNAi therapeutics: an update on delivery, Curr Opin.Mol Ther 2008; 10(2):158-167.

34. E. Galanis, P.A. Burch, R.L. Richardson, B. Lewis, H.C. Pitot, S. Frytak, C. Spier, E.T. Akporiaye, P.P. Peethambaram, J.S. Kaur, S.H. Okuno, K.K. Unni and J. Rubin: Intratumoral administration of a 1,2-dimyristyloxypropyl-3- dimethylhydroxyethyl ammonium bromide/dioleoylphosphatidylethanolamine formulation of the human interleukin-2 gene in the treatment of metastatic renal cell carcinoma, Cancer 2004; 101(11):2557-2566.

7 REFERENCES

35. M.E. Davis: The first targeted delivery of siRNA in humans via a self-assembling, cyclodextrin polymer-based nanoparticle: from concept to clinic, Mol Pharm 2009; 6(3):659-668.
36. S. Filleur, A. Courtin, S. Ait-Si-Ali, J. Guglielmi, C. Merle, A. Harel-Bellan, P. Clezardin and F. Cabon: SiRNA-mediated inhibition of vascular endothelial growth factor severely limits tumor resistance to antiangiogenic thrombospondin-1 and slows tumor vascularization and growth, Cancer Res. 2003; 63(14):3919-3922.
37. W.J. Kim, L.V. Christensen, S. Jo, J.W. Yockman, J.H. Jeong, Y.H. Kim and S.W. Kim: Cholesteryl Oligoarginine Delivering Vascular Endothelial Growth Factor siRNA Effectively Inhibits Tumor Growth in Colon Adenocarcinoma, Mol.Ther. 2006; 14(3):343-350.
38. M. Gunther, E. Wagner and M. Ogris: Specific targets in tumor tissue for the delivery of therapeutic genes, Curr.Med.Chem Anticancer Agents 2005; 5(2):157-171.
39. F. Sakurai, T. Terada, M. Maruyama, Y. Watanabe, F. Yamashita, Y. Takakura and M. Hashida: Therapeutic effect of intravenous delivery of lipoplexes containing the interferon-beta gene and poly I: poly C in a murine lung metastasis model, Cancer Gene Ther. 2003; 10(9):661-668.
40. N. Tietze, J. Pelisek, A. Philipp, W. Roedl, T. Merdan, P. Tarcha, M. Ogris and E. Wagner: Induction of apoptosis in murine neuroblastoma by systemic delivery of transferrin-shielded siRNA polyplexes for downregulation of Ran, Oligonucleotides 2008; 18(2):161-174.
41. A.S. Rait, K.F. Pirollo, L. Xiang, D. Ulick and E.H. Chang: Tumor-targeting, systemically delivered antisense HER-2 chemosensitizes human breast cancer xenografts irrespective of HER-2 levels, Mol Med 2002; 8(8):475-486.
42. A. Aigner, D. Fischer, T. Merdan, C. Brus, T. Kissel and F. Czubayko: Delivery of unmodified bioactive ribozymes by an RNA-stabilizing polyethylenimine (LMW-PEI) efficiently down-regulates gene expression, Gene Ther 2002; 9(24):1700-1707.
43. B. Urban-Klein, S. Werth, S. Abuharbeid, F. Czubayko and A. Aigner: RNAi-mediated gene-targeting through systemic application of polyethylenimine (PEI)-complexed siRNA in vivo, Gene Ther. 2005; 12(5):461-466.
44. T. Merdan, J. Kopecek and T. Kissel: Prospects for cationic polymers in gene and oligonucleotide therapy against cancer, Adv.Drug Deliv.Rev. 2002; 54(5):715-758.
45. T.G. Park, J.H. Jeong and S.W. Kim: Current status of polymeric gene delivery systems, Adv.Drug Deliv.Rev. 2006; 58(4):467-486.
46. I.R. Gilmore, S.P. Fox, A.J. Hollins and S. Akhtar: Delivery strategies for siRNA-mediated gene silencing, Curr.Drug Deliv. 2006; 3(2):147-145.
47. Y.K. Oh and T.G. Park: siRNA delivery systems for cancer treatment, Adv Drug Deliv Rev 2009; 61(10):850-862.
48. C.A. Lipinski: Drug-like properties and the causes of poor solubility and poor permeability, J Pharmacol Toxicol Methods 2000; 44(1):235-249.
49. J.A. Wolff and V. Budker: The mechanism of naked DNA uptake and expression, Adv Genet 2005; 543-20.
50. J.A. Wolff, R.W. Malone, P. Williams, W. Chong, G. Acsadi, A. Jani and P.L. Felgner: Direct gene transfer into mouse muscle in vivo, Science 1990; 247(4949 Pt 1):1465-1468.
51. F. Liu, Y. Song and D. Liu: Hydrodynamics-based transfection in animals by systemic administration of plasmid DNA, Gene Ther 1999; 6(7):1258-1266.

7 REFERENCES

52. K.S. Kim, D.S. Kim, K.H. Chung and Y.S. Park: Inhibition of angiogenesis and tumor progression by hydrodynamic cotransfection of angiostatin K1-3, endostatin, and saxatilin genes, Cancer Gene Ther 2006; 13(6):563-571.

53. H. Yazawa, T. Murakami, H.M. Li, T. Back, K. Kurosaka, Y. Suzuki, L. Shorts, Y. Akiyama, K. Maruyama, E. Parsoneault, R.H. Wiltrout and M. Watanabe: Hydrodynamics-based gene delivery of naked DNA encoding fetal liver kinase-1 gene effectively suppresses the growth of pre-existing tumors, Cancer Gene Ther 2006; 13(11):993-1001.

54. S.J. Reich, J. Fosnot, A. Kuroki, W. Tang, X. Yang, A.M. Maguire, J. Bennett and M.J. Tolentino: Small interfering RNA (siRNA) targeting VEGF effectively inhibits ocular neovascularization in a mouse model, Mol Vis 2003; 9210-216.

55. M.J. Tolentino, A.J. Brucker, J. Fosnot, G.S. Ying, I.H. Wu, G. Malik, S. Wan and S.J. Reich: Intravitreal injection of vascular endothelial growth factor small interfering RNA inhibits growth and leakage in a nonhuman primate, laser-induced model of choroidal neovascularization, Retina 2004; 24(4):660.

56. http://clinicaltrials.gov.

57. R.I. Mahato, K. Kawabata, Y. Takakura and M. Hashida: In vivo disposition characteristics of plasmid DNA complexed with cationic liposomes, J Drug Target 1995; 3(2):149-157.

58. K. Kawabata, Y. Takakura and M. Hashida: The fate of plasmid DNA after intravenous injection in mice: involvement of scavenger receptors in its hepatic uptake, Pharm Res 1995; 12(6):825-830.

59. B.E. Houk, G. Hochhaus and J.A. Hughes: Kinetic modeling of plasmid DNA degradation in rat plasma, AAPS PharmSci 1999; 1(3):E9.

60. B.E. Houk, R. Martin, G. Hochhaus and J.A. Hughes: Pharmacokinetics of plasmid DNA in the rat, Pharm Res 2001; 18(1):67-74.

61. L. Jager and A. Ehrhardt: Emerging adenoviral vectors for stable correction of genetic disorders, Curr Gene Ther 2007; 7(4):272-283.

62. M. Cavazzana-Calvo and A. Fischer: Gene therapy for severe combined immunodeficiency: are we there yet?, J Clin.Invest 2007; 117(6):1456-1465.

63. Y. Yi, S.H. Hahm and K.H. Lee: Retroviral gene therapy: safety issues and possible solutions, Curr Gene Ther 2005; 5(1):25-35.

64. V. Nair: Retrovirus-induced oncogenesis and safety of retroviral vectors, Curr Opin Mol Ther 2008; 10(5):431-438.

65. G.L. Odom, P. Gregorevic and J.S. Chamberlain: Viral-mediated gene therapy for the muscular dystrophies: successes, limitations and recent advances, Biochim.Biophys.Acta 2007; 1772(2):243-262.

66. E. Check: Regulators split on gene therapy as patient shows signs of cancer, Nature 2002; 419(6907):545-546.

67. Z. Li, J. Dullmann, B. Schiedlmeier, M. Schmidt, C. von Kalle, J. Meyer, M. Forster, C. Stocking, A. Wahlers, O. Frank, W. Ostertag, K. Kuhlcke, H.G. Eckert, B. Fehse and C. Baum: Murine leukemia induced by retroviral gene marking, Science 2002; 296(5567):497.

68. G. Zuber, E. Dauty, M. Nothisen, P. Belguise and J.P. Behr: Towards synthetic viruses, Adv Drug Deliv Rev 2001; 52(3):245-253.

7 REFERENCES

69. T. Niidome and L. Huang: Gene therapy progress and prospects: nonviral vectors, Gene Ther 2002; 9(24):1647-1652.

70. E. Wagner: Strategies to improve DNA polyplexes for in vivo gene transfer: will "artificial viruses" be the answer?, Pharm.Res. 2004; 21(1):8-14.

71. S.D. Li and L. Huang: Non-viral is superior to viral gene delivery, J.Control Release 2007; 123(3):181-183.

72. D. Liu, T. Ren and X. Gao: Cationic transfection lipids, Curr.Med Chem. 2003; 10(14):1307-1315.

73. N. Duzgunes, C.T. De Ilarduya, S. Simoes, R.I. Zhdanov, K. Konopka and M.C. Pedroso de Lima: Cationic liposomes for gene delivery: novel cationic lipids and enhancement by proteins and peptides, Curr.Med.Chem. 2003; 10(14):1213-1220.

74. S.L. Hart: Lipid carriers for gene therapy, Curr.Drug Deliv. 2005; 2(4):423-428.

75. W. Li and F.C. Szoka, Jr.: Lipid-based nanoparticles for nucleic acid delivery, Pharm.Res 2007; 24(3):438-449.

76. S.C. De Smedt, J. Demeester and W.E. Hennink: Cationic polymer based gene delivery systems, Pharm.Res. 2000; 17(2):113-126.

77. D.W. Pack, A.S. Hoffman, S. Pun and P.S. Stayton: Design and development of polymers for gene delivery, Nat.Rev.Drug Discov. 2005; 4(7):581-593.

78. M.J. Tiera, F.O. Winnik and J.C. Fernandes: Synthetic and natural polycations for gene therapy: state of the art and new perspectives, Curr.Gene Ther. 2006; 6(1):59-71.

79. D. Schaffert and E. Wagner: Gene therapy progress and prospects: synthetic polymer-based systems, Gene Ther 2008; 15(16):1131-1138.

80. E. Wagner, M. Cotten, R. Foisner and M.L. Birnstiel: Transferrin-polycation-DNA complexes: the effect of polycations on the structure of the complex and DNA delivery to cells, Proc.Natl.Acad.Sci.U.S.A 1991; 88(10):4255-4259.

81. V. Bulmus, M. Woodward, L. Lin, N. Murthy, P. Stayton and A. Hoffman: A new pH-responsive and glutathione-reactive, endosomal membrane-disruptive polymeric carrier for intracellular delivery of biomolecular drugs, J.Control Release 2003; 93(2):105-120.

82. J. Soutschek, A. Akinc, B. Bramlage, K. Charisse, R. Constien, M. Donoghue, S. Elbashir, A. Geick, P. Hadwiger, J. Harborth, M. John, V. Kesavan, G. Lavine, R.K. Pandey, T. Racie, K.G. Rajeev, I. Rohl, I. Toudjarska, G. Wang, S. Wuschko, D. Bumcrot, K. Koteliansky, S. Limmer, M. Manoharan and H.P. Vornlocher: Therapeutic silencing of an endogenous gene by systemic administration of modified siRNAs, Nature 2004; 432(7014):173-178.

83. A.M. Derfus, A.A. Chen, D.H. Min, E. Ruoslahti and S.N. Bhatia: Targeted quantum dot conjugates for siRNA delivery, Bioconjug Chem 2007; 18(5):1391-1396.

84. D.B. Rozema, D.L. Lewis, D.H. Wakefield, S.C. Wong, J.J. Klein, P.L. Roesch, S.L. Bertin, T.W. Reppen, Q. Chu, A.V. Blokhin, J.E. Hagstrom and J.A. Wolff: Dynamic PolyConjugates for targeted in vivo delivery of siRNA to hepatocytes, Proc.Natl.Acad.Sci U.S.A 2007; 104(32):12982-12987.

85. M. Meyer, C. Dohmen, A. Philipp, D. Kiener, G. Maiwald, G. Scheu, M. Ogris and E. Wagner: Synthesis and Biological Evaluation of a Bioresponsive and Endosomolytic siRNA-Polymer Conjugate, Mol Pharm 2009; 6(3):752-762.

86. K.A. Mislick and J.D. Baldeschwieler: Evidence for the role of proteoglycans in cation-mediated gene transfer, Proc.Natl.Acad.Sci.U.S.A 1996; 93(22):12349-12354.

7 REFERENCES

87. I. Kopatz, J.S. Remy and J.P. Behr: A model for non-viral gene delivery: through syndecan adhesion molecules and powered by actin, J Gene Med 2004; 6(7):769-776.

88. P.L. Felgner, Y. Barenholz, J.P. Behr, S.H. Cheng, P. Cullis, L. Huang, J.A. Jessee, L. Seymour, F. Szoka, A.R. Thierry, E. Wagner and G. Wu: Nomenclature for synthetic gene delivery systems, Hum Gene Ther 1997; 8(5):511-512.

89. R.J. Lee and L. Huang: Lipidic vector systems for gene transfer, Crit Rev.Ther Drug Carrier Syst. 1997; 14(2):173-206.

90. M.C. Pedroso de Lima, S. Simoes, P. Pires, H. Faneca and N. Duzgunes: Cationic lipid-DNA complexes in gene delivery: from biophysics to biological applications, Adv Drug Deliv Rev 2001; 47(2-3):277-294.

91. W. Li, Z. Huang, J.A. MacKay, S. Grube and F.C. Szoka, Jr.: Low-pH-sensitive poly(ethylene glycol) (PEG)-stabilized plasmid nanolipoparticles: effects of PEG chain length, lipid composition and assembly conditions on gene delivery, J Gene Med 2005; 7(1):67-79.

92. J. Lee and B.R. Lentz: Evolution of lipidic structures during model membrane fusion and the relation of this process to cell membrane fusion, Biochemistry 1997; 36(21):6251-6259.

93. R. Wattiaux, M. Jadot, M.T. Warnier-Pirotte and S. Wattiaux-De Coninck: Cationic lipids destabilize lysosomal membrane in vitro, FEBS Lett 1997; 417(2):199-202.

94. I.M. Hafez, N. Maurer and P.R. Cullis: On the mechanism whereby cationic lipids promote intracellular delivery of polynucleic acids, Gene Ther 2001; 8(15):1188-1196.

95. I.S. Zuhorn and D. Hoekstra: On the mechanism of cationic amphiphile-mediated transfection. To fuse or not to fuse: is that the question?, J.Membr.Biol. 2002; 189(3):167-179.

96. D. Hoekstra, J. Rejman, L. Wasungu, F. Shi and I. Zuhorn: Gene delivery by cationic lipids: in and out of an endosome, Biochem.Soc Trans. 2007; 35(Pt 1):68-71.

97. Y. Xu and F.C. Szoka, Jr.: Mechanism of DNA release from cationic liposome/DNA complexes used in cell transfection, Biochemistry 1996; 35(18):5616-5623.

98. O. Zelphati and F.C. Szoka: Mechanism of oligonucleotide release from cationic liposomes, Proc.Natl.Acad.Sci.U.S.A. 1996; 93(0027-8424):11493-11498.

99. D.J. Gary, N. Puri and Y.Y. Won: Polymer-based siRNA delivery: perspectives on the fundamental and phenomenological distinctions from polymer-based DNA delivery, J Control Release 2007; 121(1-2):64-73.

100. O. Boussif, F. Lezoualc'h, M.A. Zanta, M.D. Mergny, D. Scherman, B. Demeneix and J.P. Behr: A versatile vector for gene and oligonucleotide transfer into cells in culture and in vivo: polyethylenimine, Proc.Natl.Acad.Sci.U.S.A 1995; 92(16):7297-7301.

101. D. Goula, C. Benoist, S. Mantero, G. Merlo, G. Levi and B.A. Demeneix: Polyethylenimine-based intravenous delivery of transgenes to mouse lung, Gene Ther 1998; 5(9):1291-1295.

102. W.T. Godbey, K.K. Wu and A.G. Mikos: Poly(ethylenimine) and its role in gene delivery 988, J.Control Release 1999; 60(2-3):149-160.

103. S.M. Zou, P. Erbacher, J.S. Remy and J.P. Behr: Systemic linear polyethylenimine (L-PEI)-mediated gene delivery in the mouse, J.Gene Med. 2000; 2(2):128-134.

104. U. Lungwitz, M. Breunig, T. Blunk and A. Gopferich: Polyethylenimine-based non-viral gene delivery systems, Eur.J.Pharm.Biopharm. 2005; 60(2):247-266.

7 REFERENCES

105. W.T. Godbey, K.K. Wu and A.G. Mikos: Tracking the intracellular path of poly(ethylenimine)/DNA complexes for gene delivery, Proc Natl Acad Sci U S A 1999; 96(9):5177-5181.

106. T. Bieber, W. Meissner, S. Kostin, A. Niemann and H.P. Elsasser: Intracellular route and transcriptional competence of polyethylenimine-DNA complexes, J.Control Release 2002; 82(2-3):441-454.

107. N.D. Sonawane, F.C. Szoka, Jr. and A.S. Verkman: Chloride Accumulation and Swelling in Endosomes Enhances DNA Transfer by Polyamine-DNA Polyplexes, J.Biol.Chem. 2003; 278(45):44826-44831.

108. A. Akinc, M. Thomas, A.M. Klibanov and R. Langer: Exploring polyethylenimine-mediated DNA transfection and the proton sponge hypothesis, J Gene Med 2005; 7(5):657-663.

109. D. Fischer, T. Bieber, Y. Li, H.P. Elsasser and T. Kissel: A novel non-viral vector for DNA delivery based on low molecular weight, branched polyethylenimine: effect of molecular weight on transfection efficiency and cytotoxicity, Pharm Res 1999; 16(8):1273-1279.

110. L. Wightman, R. Kircheis, V. Rossler, S. Carotta, R. Ruzicka, M. Kursa and E. Wagner: Different behavior of branched and linear polyethylenimine for gene delivery in vitro and in vivo, J.Gene Med. 2001; 3(4):362-372.

111. M. Breunig, U. Lungwitz, R. Liebl and A. Goepferich: Breaking up the correlation between efficacy and toxicity for nonviral gene delivery, Proc.Natl.Acad.Sci U.S.A 2007; 104(36):14454-14459.

112. P. Chollet, M.C. Favrot, A. Hurbin and J.L. Coll: Side-effects of a systemic injection of linear polyethylenimine-DNA complexes, J Gene Med. 2002; 4(1):84-91.

113. S.M. Moghimi, P. Symonds, J.C. Murray, A.C. Hunter, G. Debska and A. Szewczyk: A two-stage poly(ethylenimine)-mediated cytotoxicity: implications for gene transfer/therapy, Mol Ther 2005; 11(6):990-995.

114. S. Hong, P.R. Leroueil, E.K. Janus, J.L. Peters, M.M. Kober, M.T. Islam, B.G. Orr, J.R. Baker, Jr. and M.M. Banaszak Holl: Interaction of polycationic polymers with supported lipid bilayers and cells: nanoscale hole formation and enhanced membrane permeability, Bioconjug Chem 2006; 17(3):728-734.

115. C. Plank, K. Mechtler, F.C. Szoka, Jr. and E. Wagner: Activation of the complement system by synthetic DNA complexes: a potential barrier for intravenous gene delivery, Hum.Gene Ther 1996; 7(12):1437-1446.

116. M. Ogris, S. Brunner, S. Schuller, R. Kircheis and E. Wagner: PEGylated DNA/transferrin-PEI complexes: reduced interaction with blood components, extended circulation in blood and potential for systemic gene delivery, Gene Ther 1999; 6(4):595-605.

117. R. Kircheis, S. Schuller, S. Brunner, M. Ogris, K.H. Heider, W. Zauner and E. Wagner: Polycation-based DNA complexes for tumor-targeted gene delivery in vivo, J Gene Med 1999; 1(2):111-120.

118. A. Brownlie, I.F. Uchegbu and A.G. Schatzlein: PEI-based vesicle-polymer hybrid gene delivery system with improved biocompatibility, Int.J Pharm. 2004; 274(1-2):41-52.

119. M.A. Gosselin, W. Guo and R.J. Lee: Efficient gene transfer using reversibly cross-linked low molecular weight polyethylenimine, Bioconjug Chem 2001; 12(6):989-994.

7 REFERENCES

120. M.L. Forrest, J.T. Koerber and D.W. Pack: A degradable polyethylenimine derivative with low toxicity for highly efficient gene delivery, Bioconjug.Chem. 2003; 14(5):934-940.

121. C.C. Lee, J.A. MacKay, J.M. Frechet and F.C. Szoka: Designing dendrimers for biological applications, Nat.Biotechnol. 2005; 23(12):1517-1526.

122. H.K. de Wolf, J. Luten, C.J. Snel, C. Oussoren, W.E. Hennink and G. Storm: In vivo tumor transfection mediated by polyplexes based on biodegradable poly(DMAEA)-phosphazene, J Control Release 2005; 109(1-3):275-287.

123. Y.H. Kim, J.H. Park, M. Lee, T.G. Park and S.W. Kim: Polyethylenimine with acid-labile linkages as a biodegradable gene carrier, J Control Release 2005; 103(1):209-219.

124. T.I. Kim, H.J. Seo, J.S. Choi, J.K. Yoon, J.U. Baek, K. Kim and J.S. Park: Synthesis of biodegradable cross-linked poly(beta-amino ester) for gene delivery and its modification, inducing enhanced transfection efficiency and stepwise degradation, Bioconjug.Chem. 2005; 16(5):1140-1148.

125. M. Thomas, Q. Ge, J.J. Lu, J. Chen and A.M. Klibanov: Cross-linked small polyethylenimines: while still nontoxic, deliver DNA efficiently to mammalian cells in vitro and in vivo, Pharm.Res. 2005; 22(3):373-380.

126. Z. Zhong, Y. Song, J.F. Engbersen, M.C. Lok, W.E. Hennink and J. Feijen: A versatile family of degradable non-viral gene carriers based on hyperbranched poly(ester amine)s, J Control Release 2005; 109(1-3):317-329.

127. J. Kloeckner, E. Wagner and M. Ogris: Degradable gene carriers based on oligomerized polyamines, Eur.J.Pharm.Sci. 2006; 29(5):414-425.

128. J. Kloeckner, S. Bruzzano, M. Ogris and E. Wagner: Gene carriers based on hexanediol diacrylate linked oligoethylenimine: effect of chemical structure of polymer on biological properties, Bioconjug Chem 2006; 17(5):1339-1345.

129. J. Kloeckner, S. Boeckle, D. Persson, W. Roedl, M. Ogris, K. Berg and E. Wagner: DNA polyplexes based on degradable oligoethylenimine-derivatives: Combination with EGF receptor targeting and endosomal release functions, J.Control Release 2006; 116(2):115-122.

130. L.V. Christensen, C.W. Chang, W.J. Kim, S.W. Kim, Z. Zhong, C. Lin, J.F. Engbersen and J. Feijen: Reducible poly(amido ethylenimine)s designed for triggered intracellular gene delivery, Bioconjug Chem 2006; 17(5):1233-1240.

131. M. Neu, O. Germershaus, S. Mao, K.H. Voigt, M. Behe and T. Kissel: Crosslinked nanocarriers based upon poly(ethylene imine) for systemic plasmid delivery: in vitro characterization and in vivo studies in mice, J.Control Release 2007; 118(3):370-380.

132. J.J. Hoon, L.V. Christensen, J.W. Yockman, Z. Zhong, J.F. Engbersen, K.W. Jong, J. Feijen and K.S. Wan: Reducible poly(amido ethylenimine) directed to enhance RNA interference, Biomaterials 2007; 28(10):1912-1917.

133. Y. Lee, H. Mo, H. Koo, J.Y. Park, M.Y. Cho, G.W. Jin and J.S. Park: Visualization of the degradation of a disulfide polymer, linear poly(ethylenimine sulfide), for gene delivery, Bioconjug Chem 2007; 18(1):13-18.

134. V. Knorr, V. Russ, L. Allmendinger, M. Ogris and E. Wagner: Acetal linked oligoethylenimines for use as pH-sensitive gene carriers, Bioconjug Chem 2008; 19(8):1625-1634.

135. V. Russ, H. Elfberg, C. Thoma, J. Kloeckner, M. Ogris and E. Wagner: Novel degradable oligoethylenimine acrylate ester-based pseudodendrimers for in vitro and in vivo gene transfer, Gene Ther 2008; 15(1):18-29.

136. V. Russ, M. Gunther, A. Halama, M. Ogris and E. Wagner: Oligoethylenimine-grafted polypropylenimine dendrimers as degradable and biocompatible synthetic vectors for gene delivery, J Control Release 2008; 132(2):131-140.

137. H.Q. Mao, K. Roy, L. Troung, K.A. Janes, K.Y. Lin, Y. Wang, J.T. August and K.W. Leong: Chitosan-DNA nanoparticles as gene carriers: synthesis, characterization and transfection efficiency, J Control Release 2001; 70(3):399-421.

138. H. Katas and H.O. Alpar: Development and characterisation of chitosan nanoparticles for siRNA delivery, J Control Release 2006; 115(2):216-225.

139. X. Liu, K.A. Howard, M. Dong, M.O. Andersen, U.L. Rahbek, M.G. Johnsen, O.C. Hansen, F. Besenbacher and J. Kjems: The influence of polymeric properties on chitosan/siRNA nanoparticle formulation and gene silencing, Biomaterials 2007; 28(6):1280-1288.

140. K.A. Howard, S.R. Paludan, M.A. Behlke, F. Besenbacher, B. Deleuran and J. Kjems: Chitosan/siRNA nanoparticle-mediated TNF-alpha knockdown in peritoneal macrophages for anti-inflammatory treatment in a murine arthritis model, Mol Ther 2009; 17(1):162-168.

141. S. Mao, W. Sun and T. Kissel: Chitosan-based formulations for delivery of DNA and siRNA, Adv Drug Deliv Rev 2009;

142. V.L. Truong-Le, S.M. Walsh, E. Schweibert, H.Q. Mao, W.B. Guggino, J.T. August and K.W. Leong: Gene transfer by DNA-gelatin nanospheres, Arch Biochem Biophys 1999; 361(1):47-56.

143. G. Kaul and M. Amiji: Tumor-targeted gene delivery using poly(ethylene glycol)-modified gelatin nanoparticles: in vitro and in vivo studies, Pharm Res 2005; 22(6):951-961.

144. K. Zwiorek, J. Kloeckner, E. Wagner and C. Coester: Gelatin nanoparticles as a new and simple gene delivery system, J Pharm Pharm Sci 2005; 7(4):22-28.

145. F. Hoffmann, G. Sass, J. Zillies, S. Zahler, G. Tiegs, A. Hartkorn, S. Fuchs, J. Wagner, G. Winter, C. Coester, A.L. Gerbes and A.M. Vollmar: A novel technique for selective NF-kappaB inhibition in Kupffer cells: contrary effects in fulminant hepatitis and ischaemia-reperfusion, Gut 2009; 58(12):1670-1678.

146. K. Iwaki, K. Shibata, M. Ohta, Y. Endo, H. Uchida, M. Tominaga, R. Okunaga, S. Kai and S. Kitano: A small interfering RNA targeting proteinase-activated receptor-2 is effective in suppression of tumor growth in a Panc1 xenograft model, Int J Cancer 2008; 122(3):658-663.

147. E. Kawata, E. Ashihara, S. Kimura, K. Takenaka, K. Sato, R. Tanaka, A. Yokota, Y. Kamitsuji, M. Takeuchi, J. Kuroda, F. Tanaka, T. Yoshikawa and T. Maekawa: Administration of PLK-1 small interfering RNA with atelocollagen prevents the growth of liver metastases of lung cancer, Mol Cancer Ther 2008; 7(9):2904-2912.

148. M. Berton, E. Allemann, C.A. Stein and R. Gurny: Highly loaded nanoparticulate carrier using an hydrophobic antisense oligonucleotide complex, Eur J Pharm Sci 1999; 9(2):163-170.

149. K.K. Sandhu, C.M. McIntosh, J.M. Simard, S.W. Smith and V.M. Rotello: Gold Nanoparticle-Mediated Transfection of Mammalian Cells 1035, Bioconjug.Chem. 2002; 13(1):3-6.

150. N. Toub, J.R. Bertrand, A. Tamaddon, H. Elhamess, H. Hillaireau, A. Maksimenko, J. Maccario, C. Malvy, E. Fattal and P. Couvreur: Efficacy of siRNA nanocapsules targeted against the EWS-Fli1 oncogene in Ewing sarcoma, Pharm.Res. 2006; 23(5):892-900.

7 REFERENCES

151. S.H. Kim, J.H. Jeong, S.H. Lee, S.W. Kim and T.G. Park: LHRH receptor-mediated delivery of siRNA using polyelectrolyte complex micelles self-assembled from siRNA-PEG-LHRH conjugate and PEI, Bioconjug Chem 2008; 19(11):2156-2162.
152. C.M. Wiethoff and C.R. Middaugh: Barriers to nonviral gene delivery, J Pharm.Sci 2003; 92(2):203-217.
153. M. Ogris, G. Walker, T. Blessing, R. Kircheis, M. Wolschek and E. Wagner: Tumor-targeted gene therapy: strategies for the preparation of ligand-polyethylene glycol-polyethylenimine/DNA complexes, J.Control Release 2003; 91(1-2):173-181.
154. E. Ambegia, S. Ansell, P. Cullis, J. Heyes, L. Palmer and I. MacLachlan: Stabilized plasmid-lipid particles containing PEG-diacylglycerols exhibit extended circulation lifetimes and tumor selective gene expression, Biochim.Biophys Acta 2005; 1669(2):155-163.
155. L. Peeters, N.N. Sanders, A. Jones, J. Demeester and S.C. De Smedt: Post-pegylated lipoplexes are promising vehicles for gene delivery in RPE cells, J Control Release 2007; 121(3):208-217.
156. X. Zhang, S.R. Pan, H.M. Hu, G.F. Wu, M. Feng, W. Zhang and X. Luo: Poly(ethylene glycol)-block-polyethylenimine copolymers as carriers for gene delivery: effects of PEG molecular weight and PEGylation degree, J Biomed Mater Res A 2008; 84(3):795-804.
157. R.S. Burke and S.H. Pun: Extracellular Barriers to in Vivo PEI and PEGylated PEI Polyplex-Mediated Gene Delivery to the Liver, Bioconjug Chem 2008; 19(3):693-704.
158. K. Buyens, B. Lucas, K. Raemdonck, K. Braeckmans, J. Vercammen, J. Hendrix, Y. Engelborghs, S.C. De Smedt and N.N. Sanders: A fast and sensitive method for measuring the integrity of siRNA-carrier complexes in full human serum, J Control Release 2008; 126(1):67-76.
159. A.L. Bolcato-Bellemin, M.E. Bonnet, G. Creusat, P. Erbacher and J.P. Behr: Sticky overhangs enhance siRNA-mediated gene silencing, Proc.Natl.Acad.Sci U.S.A 2007; 104(41):16050-16055.
160. A. Zintchenko, A. Philipp, A. Dehshahri and E. Wagner: Simple Modifications of Branched PEI Lead to Highly Efficient siRNA Carriers with Low Toxicity, Bioconjug Chem 2008; 19(7):1448-1455.
161. D. Oupicky, R.C. Carlisle and L.W. Seymour: Triggered intracellular activation of disulfide crosslinked polyelectrolyte gene delivery complexes with extended systemic circulation in vivo, Gene Ther 2001; 8(9):713-724.
162. M. Neu, J. Sitterberg, U. Bakowsky and T. Kissel: Stabilized nanocarriers for plasmids based upon cross-linked poly(ethylene imine), Biomacromolecules. 2006; 7(12):3428-3438.
163. M. Neu, O. Germershaus, M. Behe and T. Kissel: Bioreversibly crosslinked polyplexes of PEI and high molecular weight PEG show extended circulation times in vivo, J.Control Release 2007; 124(1-2):69-80.
164. V. Russ and E. Wagner: Cell and Tissue Targeting of Nucleic Acids for Cancer Gene Therapy, Pharm.Res. 2007; 24(6):1047-1057.
165. A. Philipp, M. Meyer and E. Wagner: Extracellular targeting of synthetic therapeutic nucleic Acid formulations, Curr Gene Ther 2008; 8(5):324-334.
166. J. Suh, D. Wirtz and J. Hanes: Efficient active transport of gene nanocarriers to the cell nucleus, Proc.Natl.Acad.Sci.U.S.A 2003; 100(7):3878-3882.
167. S.S. Davis: Biomedical applications of nanotechnology--implications for drug targeting and gene therapy 1008, Trends Biotechnol. 1997; 15(6):217-224.

7 REFERENCES

168. M.G. Lampugnani and E. Dejana: Interendothelial junctions: structure, signalling and functional roles, Curr.Opin.Cell Biol 1997; 9(5):674-682.

169. S. Boeckle and E. Wagner: Optimizing targeted gene delivery: chemical modification of viral vectors and synthesis of artificial virus vector systems, AAPS.J. 2006; 8(4):E731-E742.

170. E. Mastrobattista, M.A. van der Aa, W.E. Hennink and D.J. Crommelin: Artificial viruses: a nanotechnological approach to gene delivery, Nat.Rev.Drug Discov. 2006; 5(2):115-121.

171. E. Wagner: Converging Paths of Viral and Non-viral Vector Engineering, Mol Ther 2008; 16(1):1-2.

172. D. Oupicky and V. Diwadkar: Stimuli-responsive gene delivery vectors, Curr.Opin.Mol Ther 2003; 5(4):345-350.

173. M. Kursa, G.F. Walker, V. Roessler, M. Ogris, W. Roedl, R. Kircheis and E. Wagner: Novel Shielded Transferrin-Polyethylene Glycol-Polyethylenimine/DNA Complexes for Systemic Tumor-Targeted Gene Transfer, Bioconjug.Chem. 2003; 14(1):222-231.

174. M. Meyer and E. Wagner: pH-responsive shielding of non-viral gene vectors, Expert.Opin.Drug Deliv. 2006; 3(5):563-571.

175. P.R. Dash, M.L. Read, K.D. Fisher, K.A. Howard, M. Wolfert, D. Oupicky, V. Subr, J. Strohalm, K. Ulbrich and L.W. Seymour: Decreased binding to proteins and cells of polymeric gene delivery vectors surface modified with a multivalent hydrophilic polymer and retargeting through attachment of transferrin 961, J.Biol.Chem. 2000; 275(6):3793-3802.

176. D. Oupicky, M. Ogris and L.W. Seymour: Development of long-circulating polyelectrolyte complexes for systemic delivery of genes, J Drug Target 2002; 10(2):93-98.

177. P. Lemieux, N. Guerin, G. Paradis, R. Proulx, L. Chistyakova, A. Kabanov and V. Alakhov: A combination of poloxamers increases gene expression of plasmid DNA in skeletal muscle 983, Gene Ther. 2000; 7(11):986-991.

178. H. Maeda: The enhanced permeability and retention (EPR) effect in tumor vasculature: the key role of tumor-selective macromolecular drug targeting, Adv Enzyme Regul 2001; 41189-207.

179. M.F. Wolschek, C. Thallinger, M. Kursa, V. Rossler, M. Allen, C. Lichtenberger, R. Kircheis, T. Lucas, M. Willheim, W. Reinisch, A. Gangl, E. Wagner and B. Jansen: Specific systemic nonviral gene delivery to human hepatocellular carcinoma xenografts in SCID mice, Hepatology 2002; 36(5):1106-1114.

180. S. Mao, M. Neu, O. Germershaus, O. Merkel, J. Sitterberg, U. Bakowsky and T. Kissel: Influence of polyethylene glycol chain length on the physicochemical and biological properties of poly(ethylene imine)-graft-poly(ethylene glycol) block copolymer/siRNA polyplexes, Bioconjug.Chem. 2006; 17(5):1209-1218.

181. C.W. Beh, W.Y. Seow, Y. Wang, Y. Zhang, Z.Y. Ong, P.L. Ee and Y.Y. Yang: Efficient delivery of Bcl-2-targeted siRNA using cationic polymer nanoparticles: downregulating mRNA expression level and sensitizing cancer cells to anticancer drug, Biomacromolecules 2009; 10(1):41-48.

182. F.J. Verbaan, C. Oussoren, C.J. Snel, D.J. Crommelin, W.E. Hennink and G. Storm: Steric stabilization of poly(2-(dimethylamino)ethyl methacrylate)-based polyplexes mediates prolonged circulation and tumor targeting in mice, J Gene Med 2004; 6(1):64-75.

7 REFERENCES

183. O. Taratula, O.B. Garbuzenko, P. Kirkpatrick, I. Pandya, R. Savla, V.P. Pozharov, H. He and T. Minko: Surface-engineered targeted PPI dendrimer for efficient intracellular and intratumoral siRNA delivery, J Control Release 2009; 140(3):284-293.

184. G.Y. Wu and C.H. Wu: Receptor-mediated in vitro gene transformation by a soluble DNA carrier system, J Biol Chem 1987; 262(10):4429-4432.

185. G.Y. Wu and C.H. Wu: Receptor-mediated gene delivery and expression in vivo, J Biol Chem 1988; 263(29):14621-14624.

186. E. Wagner, M. Ogris and W. Zauner: Polylysine-based transfection systems utilizing receptor-mediated delivery, Adv.Drug Deliv.Rev. 1998; 30(1-3):97-113.

187. D.V. Schaffer and D.A. Lauffenburger: Targeted synthetic gene delivery vectors, Curr.Opin.Mol Ther 2000; 2(2):155-161.

188. A.G. Schatzlein: Targeting of Synthetic Gene Delivery Systems, J Biomed.Biotechnol. 2003; 2003(2):149-158.

189. E. Wagner, C. Culmsee and S. Boeckle: Targeting of Polyplexes: Toward Synthetic Virus Vector Systems, Adv Genet 2005; 53PA333-354.

190. K.C. Wood, S.M. Azarin, W. Arap, R. Pasqualini, R. Langer and P.T. Hammond: Tumor-targeted gene delivery using molecularly engineered hybrid polymers functionalized with a tumor-homing peptide, Bioconjug Chem 2008; 19(2):403-405.

191. E. Wagner, M. Zenke, M. Cotten, H. Beug and M.L. Birnstiel: Transferrin-polycation conjugates as carriers for DNA uptake into cells, Proc.Natl.Acad.Sci.U.S.A 1990; 87(9):3410-3414.

192. H. Li and Z.M. Qian: Transferrin/transferrin receptor-mediated drug delivery, Med.Res.Rev. 2002; 22(3):225-250.

193. S. Hu-Lieskovan, J.D. Heidel, D.W. Bartlett, M.E. Davis and T.J. Triche: Sequence-specific knockdown of EWS-FLI1 by targeted, nonviral delivery of small interfering RNA inhibits tumor growth in a murine model of metastatic Ewing's sarcoma, Cancer Res. 2005; 65(19):8984-8992.

194. R. Kircheis, L. Wightman, A. Schreiber, B. Robitza, V. Rossler, M. Kursa and E. Wagner: Polyethylenimine/DNA complexes shielded by transferrin target gene expression to tumors after systemic application, Gene Ther 2001; 8(1):28-40.

195. I.J. Hildebrandt, M. Iyer, E. Wagner and S.S. Gambhir: Optical imaging of transferrin targeted PEI/DNA complexes in living subjects, Gene Ther. 2003; 10(9):758-764.

196. M.T. da Cruz, A.L. Cardoso, L.P. de Almeida, S. Simoes and M.C. de Lima: Tf-lipoplex-mediated NGF gene transfer to the CNS: neuronal protection and recovery in an excitotoxic model of brain injury, Gene Ther. 2005; 12(16):1242-1252.

197. D.W. Bartlett and M.E. Davis: Insights into the kinetics of siRNA-mediated gene silencing from live-cell and live-animal bioluminescent imaging, Nucleic Acids Res 2006; 34(1):322-333.

198. D.W. Bartlett, H. Su, I.J. Hildebrandt, W.A. Weber and M.E. Davis: Impact of tumor-specific targeting on the biodistribution and efficacy of siRNA nanoparticles measured by multimodality in vivo imaging, Proc.Natl.Acad.Sci U.S.A 2007; 104(39):15549-15554.

199. D.W. Bartlett and M.E. Davis: Impact of tumor-specific targeting and dosing schedule on tumor growth inhibition after intravenous administration of siRNA-containing nanoparticles, Biotechnol.Bioeng. 2008; 99(4):975-985.

7 REFERENCES

200. T. Blessing, M. Kursa, R. Holzhauser, R. Kircheis and E. Wagner: Different strategies for formation of pegylated EGF-conjugated PEI/DNA complexes for targeted gene delivery, Bioconjug Chem 2001; 12(4):529-537.

201. K. von Gersdorff, M. Ogris and E. Wagner: Cryoconserved shielded and EGF receptor targeted DNA polyplexes: cellular mechanisms, Eur.J Pharm.Biopharm. 2005; 60(2):279-285.

202. K. de Bruin, N. Ruthardt, K. von Gersdorff, R. Bausinger, E. Wagner, M. Ogris and C. Brauchle: Cellular dynamics of EGF receptor-targeted synthetic viruses, Mol Ther 2007; 15(7):1297-1305.

203. K. Kunath, T. Merdan, O. Hegener, H. Haberlein and T. Kissel: Integrin targeting using RGD-PEI conjugates for in vitro gene transfer, J.Gene Med. 2003; 5(7):588-599.

204. R.M. Schiffelers, A. Ansari, J. Xu, Q. Zhou, Q. Tang, G. Storm, G. Molema, P.Y. Lu, P.V. Scaria and M.C. Woodle: Cancer siRNA therapy by tumor selective delivery with ligand-targeted sterically stabilized nanoparticle, Nucleic Acids Res. 2004; 32(19):e149.

205. Y. Ikeda and K. Taira: Ligand-targeted delivery of therapeutic siRNA, Pharm.Res 2006; 23(8):1631-1640.

206. M. Oba, S. Fukushima, N. Kanayama, K. Aoyagi, N. Nishiyama, H. Koyama and K. Kataoka: Cyclic RGD peptide-conjugated polyplex micelles as a targetable gene delivery system directed to cells possessing alphavbeta3 and alphavbeta5 integrins, Bioconjug Chem 2007; 18(5):1415-1423.

207. S.H. Kim, H. Mok, J.H. Jeong, S.W. Kim and T.G. Park: Comparative evaluation of target-specific GFP gene silencing efficiencies for antisense ODN, synthetic siRNA, and siRNA plasmid complexed with PEI-PEG-FOL conjugate, Bioconjug Chem 2006; 17(1):241-244.

208. C.K. Chul, J.J. Hoon, C.H. Jung, C.O. Joe, K.S. Wan and P.T. Gwan: Folate receptor-mediated intracellular delivery of recombinant caspase-3 for inducing apoptosis, J.Control Release 2005; 108(1):121-131.

209. B. Liang, M.L. He, Z.P. Xiao, Y. Li, C.Y. Chan, H.F. Kung, X.T. Shuai and Y. Peng: Synthesis and characterization of folate-PEG-grafted-hyperbranched-PEI for tumor-targeted gene delivery, Biochem.Biophys.Res Commun. 2008; 367(4):874-880.

210. R. Kircheis, A. Kichler, G. Wallner, M. Kursa, M. Ogris, T. Felzmann, M. Buchberger and E. Wagner: Coupling of cell-binding ligands to polyethylenimine for targeted gene delivery, Gene Ther. 1997; 4(5):409-418.

211. L. Xu, C.C. Huang, W. Huang, W.H. Tang, A. Rait, Y.Z. Yin, I. Cruz, L.M. Xiang, K.F. Pirollo and E.H. Chang: Systemic tumor-targeted gene delivery by anti-transferrin receptor scFv-immunoliposomes, Mol.Cancer Ther. 2002; 1(5):337-346.

212. J.H. Jeong, M. Lee, W.J. Kim, J.W. Yockman, T.G. Park, Y.H. Kim and S.W. Kim: Anti-GAD antibody targeted non-viral gene delivery to islet beta cells, J.Control Release 2005; 107(3):562-570.

213. E. Song, P. Zhu, S.K. Lee, D. Chowdhury, S. Kussman, D.M. Dykxhoorn, Y. Feng, D. Palliser, D.B. Weiner, P. Shankar, W.A. Marasco and J. Lieberman: Antibody mediated in vivo delivery of small interfering RNAs via cell-surface receptors, Nat.Biotechnol. 2005; 23(6):709-717.

214. S. Moffatt, C. Papasakelariou, S. Wiehle and R. Cristiano: Successful in vivo tumor targeting of prostate-specific membrane antigen with a highly efficient J591/PEI/DNA molecular conjugate, Gene Ther. 2006; 13(9):761-772.

7 REFERENCES

215. K.F. Pirollo, G. Zon, A. Rait, Q. Zhou, W. Yu, R. Hogrefe and E.H. Chang: Tumor-targeting nanoimmunoliposome complex for short interfering RNA delivery, Hum.Gene Ther. 2006; 17(1):117-124.

216. B. Spankuch, I. Steinhauser, H. Wartlick, E. Kurunci-Csacsko, K.I. Strebhardt and K. Langer: Downregulation of Plk1 expression by receptor-mediated uptake of antisense oligonucleotide-loaded nanoparticles, Neoplasia. 2008; 10(3):223-234.

217. S. Mishra, P. Webster and M.E. Davis: PEGylation significantly affects cellular uptake and intracellular trafficking of non-viral gene delivery particles, Eur.J Cell Biol. 2004; 83(3):97-111.

218. P. Midoux and M. Monsigny: Efficient gene transfer by histidylated polylysine/pDNA complexes, Bioconjug.Chem. 1999; 10(3):406-411.

219. D. Putnam, C.A. Gentry, D.W. Pack and R. Langer: Polymer-based gene delivery with low cytotoxicity by a unique balance of side-chain termini, Proc Natl Acad Sci U S A 2001; 98(3):1200-1205.

220. M. Bello Roufai and P. Midoux: Histidylated polylysine as DNA vector: elevation of the imidazole protonation and reduced cellular uptake without change in the polyfection efficiency of serum stabilized negative polyplexes, Bioconjug Chem 2001; 12(1):92-99.

221. A. Kichler, C. Leborgne, J. Marz, O. Danos and B. Bechinger: Histidine-rich amphipathic peptide antibiotics promote efficient delivery of DNA into mammalian cells, Proc.Natl.Acad.Sci.U.S.A 2003; 100(4):1564-1568.

222. M.L. Read, S. Singh, Z. Ahmed, M. Stevenson, S.S. Briggs, D. Oupicky, L.B. Barrett, R. Spice, M. Kendall, M. Berry, J.A. Preece, A. Logan and L.W. Seymour: A versatile reducible polycation-based system for efficient delivery of a broad range of nucleic acids, Nucleic Acids Res. 2005; 33(9):e86.

223. M. Stevenson, V. Ramos-Perez, S. Singh, M. Soliman, J.A. Preece, S.S. Briggs, M.L. Read and L.W. Seymour: Delivery of siRNA mediated by histidine-containing reducible polycations, J Control Release 2008; 130(1):46-56.

224. E. Wagner, C. Plank, K. Zatloukal, M. Cotten and M.L. Birnstiel: Influenza virus hemagglutinin HA-2 N-terminal fusogenic peptides augment gene transfer by transferrin-polylysine-DNA complexes: toward a synthetic virus-like gene-transfer vehicle, Proc.Natl.Acad.Sci.U.S.A 1992; 89(17):7934-7938.

225. C. Plank, W. Zauner and E. Wagner: Application of membrane-active peptides for drug and gene delivery across cellular membranes, Adv.Drug Deliv.Rev. 1998; 34(1):21-35.

226. E. Wagner: Application of membrane-active peptides for nonviral gene delivery, Adv Drug Deliv.Rev 1999; 38(3):279-289.

227. S. Dramsi and P. Cossart: Listeriolysin O: a genuine cytolysin optimized for an intracellular parasite, J Cell Biol 2002; 156(6):943-946.

228. E. Wagner, K. Zatloukal, M. Cotten, H. Kirlappos, K. Mechtler, D.T. Curiel and M.L. Birnstiel: Coupling of adenovirus to transferrin-polylysine/DNA complexes greatly enhances receptor-mediated gene delivery and expression of transfected genes, Proc.Natl.Acad.Sci.U.S.A 1992; 89(13):6099-6103.

229. C.M. Wiethoff, H. Wodrich, L. Gerace and G.R. Nemerow: Adenovirus protein VI mediates membrane disruption following capsid disassembly, J Virol. 2005; 79(4):1992-2000.

230. R.W. Ruigrok, N.G. Wrigley, L.J. Calder, S. Cusack, S.A. Wharton, E.B. Brown and J.J. Skehel: Electron microscopy of the low pH structure of influenza virus haemagglutinin, Embo J 1986; 5(1):41-49.

7 REFERENCES

231. W. Zauner, D. Blaas, E. Kuechler and E. Wagner: Rhinovirus-mediated endosomal release of transfection complexes, J Virol. 1995; 69(2):1085-1092.

232. C. Plank, B. Oberhauser, K. Mechtler, C. Koch and E. Wagner: The influence of endosome-disruptive peptides on gene transfer using synthetic virus-like gene transfer systems, J Biol.Chem. 1994; 269(17):12918-12924.

233. E.J. Kwon, J.M. Bergen and S.H. Pun: Application of an HIV gp41-derived peptide for enhanced intracellular trafficking of synthetic gene and siRNA delivery vehicles, Bioconjug Chem 2008; 19(4):920-927.

234. T.B. Wyman, F. Nicol, O. Zelphati, P.V. Scaria, C. Plank and F.C. Szoka, Jr.: Design, synthesis, and characterization of a cationic peptide that binds to nucleic acids and permeabilizes bilayers, Biochemistry 1997; 36(10):3008-3017.

235. W. Li, F. Nicol and F.C. Szoka, Jr.: GALA: a designed synthetic pH-responsive amphipathic peptide with applications in drug and gene delivery, Adv.Drug Deliv.Rev. 2004; 56(7):967-985.

236. K. Sasaki, K. Kogure, S. Chaki, Y. Nakamura, R. Moriguchi, H. Hamada, R. Danev, K. Nagayama, S. Futaki and H. Harashima: An artificial virus-like nano carrier system: enhanced endosomal escape of nanoparticles via synergistic action of pH-sensitive fusogenic peptide derivatives, Anal.Bioanal.Chem 2008; 391(8):2717-2727.

237. H. Mok and T.G. Park: Self-crosslinked and reducible fusogenic peptides for intracellular delivery of siRNA, Biopolymers 2008; 89(10):881-888.

238. J.Y. Legendre, A. Trzeciak, B. Bohrmann, U. Deuschle, E. Kitas and A. Supersaxo: Dioleoylmelittin as a novel serum-insensitive reagent for efficient transfection of mammalian cells, Bioconjug Chem 1997; 8(1):57-63.

239. M. Ogris, R.C. Carlisle, T. Bettinger and L.W. Seymour: Melittin enables efficient vesicular escape and enhanced nuclear access of nonviral gene delivery vectors, J Biol Chem 2001; 276(50):47550-47555.

240. S. Boeckle, J. Fahrmeir, W. Roedl, M. Ogris and E. Wagner: Melittin analogs with high lytic activity at endosomal pH enhance transfection with purified targeted PEI polyplexes, J.Control Release 2006; 112(2):240-248.

241. C.P. Chen, J.S. Kim, E. Steenblock, D. Liu and K.G. Rice: Gene transfer with poly-melittin peptides, Bioconjug Chem 2006; 17(4):1057-1062.

242. E. Wagner: Programmed drug delivery: nanosystems for tumor targeting, Expert.Opin.Biol Ther 2007; 7(5):587-593.

243. N. Murthy, J. Campbell, N. Fausto, A.S. Hoffman and P.S. Stayton: Design and synthesis of pH-responsive polymeric carriers that target uptake and enhance the intracellular delivery of oligonucleotides, J.Control Release 2003; 89(3):365-374.

244. K. Na, V.T. Sethuraman and Y.H. Bae: Stimuli-sensitive polymeric micelles as anticancer drug carriers, Anticancer Agents Med Chem 2006; 6(6):525-535.

245. C. Fella, G.F. Walker, M. Ogris and E. Wagner: Amine-reactive pyridylhydrazone-based PEG reagents for pH-reversible PEI polyplex shielding, Eur.J Pharm.Sci 2008; 34(4-5):309-320.

246. V. Knorr, M. Ogris and E. Wagner: An acid sensitive ketal-based polyethylene glycol-oligoethylenimine copolymer mediates improved transfection efficiency at reduced toxicity, Pharm.Res 2008; 25(12):2937-2945.

247. M. Meyer, A. Philipp, R. Oskuee, C. Schmidt and E. Wagner: Breathing life into polycations: functionalization with pH-responsive endosomolytic peptides and polyethylene glycol enables siRNA delivery, J Am Chem Soc 2008; 130(11):3272-3273.

7 REFERENCES

248. S. Matsumoto, R.J. Christie, N. Nishiyama, K. Miyata, A. Ishii, M. Oba, H. Koyama, Y. Yamasaki and K. Kataoka: Environment-responsive block copolymer micelles with a disulfide cross-linked core for enhanced siRNA delivery, Biomacromolecules 2009; 10(1):119-127.

249. N. Murthy, J. Campbell, N. Fausto, A.S. Hoffman and P.S. Stayton: Bioinspired pH-Responsive Polymers for the Intracellular Delivery of Biomolecular Drugs, Bioconjug.Chem. 2003; 14(2):412-419.

250. E.R. Gillies, A.P. Goodwin and J.M. Frechet: Acetals as pH-sensitive linkages for drug delivery, Bioconjug.Chem. 2004; 15(6):1254-1263.

251. V. Knorr, L. Allmendinger, G.F. Walker, F.F. Paintner and E. Wagner: An acetal-based PEGylation reagent for pH-sensitive shielding of DNA polyplexes, Bioconjug Chem 2007; 18(4):1218-1225.

252. A. Aissaoui, B. Martin, E. Kan, N. Oudrhiri, M. Hauchecorne, J.P. Vigneron, J.M. Lehn and P. Lehn: Novel cationic lipids incorporating an acid-sensitive acylhydrazone linker: synthesis and transfection properties, J.Med.Chem 2004; 47(21):5210-5223.

253. G.F. Walker, C. Fella, J. Pelisek, J. Fahrmeir, S. Boeckle, M. Ogris and E. Wagner: Toward synthetic viruses: endosomal pH-triggered deshielding of targeted polyplexes greatly enhances gene transfer in vitro and in vivo, Mol Ther 2005; 11(3):418-425.

254. M.P. Xiong, Y. Bae, S. Fukushima, M.L. Forrest, N. Nishiyama, K. Kataoka and G.S. Kwon: pH-Responsive Multi-PEGylated Dual Cationic Nanoparticles Enable Charge Modulations for Safe Gene Delivery, ChemMedChem. 2007; 2(9):1321-1327.

255. J. DeRouchey, C. Schmidt, G.F. Walker, C. Koch, C. Plank, E. Wagner and J.O. Radler: Monomolecular assembly of siRNA and poly(ethylene glycol)-peptide copolymers, Biomacromolecules. 2008; 9(2):724-732.

256. J.S. Choi, J.A. MacKay and F.C. Szoka, Jr.: Low-pH-sensitive PEG-stabilized plasmid-lipid nanoparticles: preparation and characterization, Bioconjug.Chem. 2003; 14(2):420-429.

257. C. Masson, M. Garinot, N. Mignet, B. Wetzer, P. Mailhe, D. Scherman and M. Bessodes: pH-sensitive PEG lipids containing orthoester linkers: new potential tools for nonviral gene delivery, J Control Release 2004; 99(3):423-434.

258. S. Lin, F. Du, Y. Wang, S. Ji, D. Liang, L. Yu and Z. Li: An acid-labile block copolymer of PDMAEMA and PEG as potential carrier for intelligent gene delivery systems, Biomacromolecules. 2008; 9(1):109-115.

259. M. Oishi, S. Sasaki, Y. Nagasaki and K. Kataoka: pH-responsive oligodeoxynucleotide (ODN)-poly(ethylene glycol) conjugate through acid-labile beta-thiopropionate linkage: preparation and polyion complex micelle formation, Biomacromolecules. 2003; 4(5):1426-1432.

260. M. Oishi, Y. Nagasaki, K. Itaka, N. Nishiyama and K. Kataoka: Lactosylated poly(ethylene glycol)-siRNA conjugate through acid-labile beta-thiopropionate linkage to construct pH-sensitive polyion complex micelles achieving enhanced gene silencing in hepatoma cells, J Am.Chem.Soc. 2005; 127(6):1624-1625.

261. M. Oishi, F. Nagatsugi, S. Sasaki, Y. Nagasaki and K. Kataoka: Smart polyion complex micelles for targeted intracellular delivery of PEGylated antisense oligonucleotides containing acid-labile linkages, Chembiochem. 2005; 6(4):718-725.

262. J. Shin, P. Shum and D.H. Thompson: Acid-triggered release via dePEGylation of DOPE liposomes containing acid-labile vinyl ether PEG-lipids, J.Control Release 2003; 91(1-2):187-200.

7 REFERENCES

263. X. Guo and F.C. Szoka, Jr.: Chemical approaches to triggerable lipid vesicles for drug and gene delivery, Acc.Chem.Res. 2003; 36(5):335-341.
264. J.A. Wolff and D.B. Rozema: Breaking the Bonds: Non-viral Vectors Become Chemically Dynamic, Mol Ther 2008; 16(1):8-15.
265. D.B. Rozema, K. Ekena, D.L. Lewis, A.G. Loomis and J.A. Wolff: Endosomolysis by Masking of a Membrane-Active Agent (EMMA) for Cytoplasmic Release of Macromolecules, Bioconjug.Chem. 2003; 14(1):51-57.
266. M. Meyer, A. Zintchenko, M. Ogris and E. Wagner: A dimethylmaleic acid-melittin-polylysine conjugate with reduced toxicity, pH-triggered endosomolytic activity and enhanced gene transfer potential, J.Gene Med. 2007; 9(9):797-805.
267. X.L. Wang, S. Ramusovic, T. Nguyen and Z.R. Lu: Novel polymerizable surfactants with pH-sensitive amphiphilicity and cell membrane disruption for efficient siRNA delivery, Bioconjug Chem 2007; 18(6):2169-2177.
268. S. Asayama, T. Sekine, H. Kawakami and S. Nagaoka: Design of aminated poly(1-vinylimidazole) for a new pH-sensitive polycation to enhance cell-specific gene delivery, Bioconjug Chem 2007; 18(5):1662-1667.
269. G. Saito, G.L. Amidon and K.D. Lee: Enhanced cytosolic delivery of plasmid DNA by a sulfhydryl-activatable listeriolysin O/protamine conjugate utilizing cellular reducing potential, Gene Ther 2003; 10(1):72-83.
270. M.L. Read, K.H. Bremner, D. Oupicky, N.K. Green, P.F. Searle and L.W. Seymour: Vectors based on reducible polycations facilitate intracellular release of nucleic acids, J.Gene Med. 2003; 5(3):232-245.
271. B. Brissault, A. Kichler, C. Guis, C. Leborgne, O. Danos and H. Cheradame: Synthesis of linear polyethylenimine derivatives for DNA transfection, Bioconjug.Chem. 2003; 14(3):581-587.
272. D. Wade, A. Boman, B. Wahlin, C.M. Drain, D. Andreu, H.G. Boman and R.B. Merrifield: All-D amino acid-containing channel-forming antibiotic peptides, Proc Natl Acad Sci U S A 1990; 87(12):4761-4765.
273. T.P. King, D. Wade, M.R. Coscia, S. Mitchell, L. Kochoumian and B. Merrifield: Structure-immunogenicity relationship of melittin, its transposed analogues, and D-melittin, J Immunol 1994; 153(3):1124-1131.
274. P.J. Tarcha, J. Pelisek, T. Merdan, J. Waters, K. Cheung, K. von Gersdorff, C. Culmsee and E. Wagner: Synthesis and characterization of chemically condensed oligoethylenimine containing beta-aminopropionamide linkages for siRNA delivery, Biomaterials 2007; 28(25):3731-3740.
275. A. Philipp, X. Zhao, P. Tarcha, E. Wagner and A. Zintchenko: Hydrophobically modified oligoethylenimines as highly efficient transfection agents for siRNA delivery, Bioconjug Chem 2009; 20(11):2055-2061.
276. Z. Hassani, G.F. Lemkine, P. Erbacher, K. Palmier, G. Alfama, C. Giovannangeli, J.P. Behr and B.A. Demeneix: Lipid-mediated siRNA delivery down-regulates exogenous gene expression in the mouse brain at picomolar levels, J.Gene Med. 2005; 7(2):198-207.
277. A.C. Grayson, A.M. Doody and D. Putnam: Biophysical and structural characterization of polyethylenimine-mediated siRNA delivery in vitro, Pharm.Res. 2006; 23(8):1868-1876.

278. S. Boeckle, K. von Gersdorff, S. van der Piepen, C. Culmsee, E. Wagner and M. Ogris: Purification of polyethylenimine polyplexes highlights the role of free polycations in gene transfer, J Gene Med 2004; 6(10):1102-1111.

279. L.E. Prevette, T.E. Kodger, T.M. Reineke and M.L. Lynch: Deciphering the role of hydrogen bonding in enhancing pDNA-polycation interactions, Langmuir 2007; 23(19):9773-9784.

280. S.E. Morgan-Lappe, L.A. Tucker, X. Huang, Q. Zhang, A.V. Sarthy, D. Zakula, L. Vernetti, M. Schurdak, J. Wang and S.W. Fesik: Identification of Ras-Related Nuclear Protein, Targeting Protein for Xenopus Kinesin-like Protein 2, and Stearoyl-CoA Desaturase 1 as Promising Cancer Targets from an RNAi-Based Screen, Cancer Res. 2007; 67(9):4390-4398.

281. H. Abe, T. Kamai, H. Shirataki, T. Oyama, K. Arai and K. Yoshida: High expression of Ran GTPase is associated with local invasion and metastasis of human clear cell renal cell carcinoma, Int.J.Cancer 2008; 122(10):2391-2397.

282. S. Sazer and M. Dasso: The ran decathlon: multiple roles of Ran, J Cell Sci 2000; 113 (Pt 7)1111-1118.

283. O.J. Gruss and I. Vernos: The mechanism of spindle assembly: functions of Ran and its target TPX2, J Cell Biol 2004; 166(7):949-955.

284. M. Ogris, P. Steinlein, M. Kursa, K. Mechtler, R. Kircheis and E. Wagner: The size of DNA/transferrin-PEI complexes is an important factor for gene expression in cultured cells, Gene Ther 1998; 5(10):1425-1433.

285. A. Kabanov, J. Zhu and V. Alakhov: Pluronic Block Copolymers for Gene Delivery, Adv Genet 2005; 53PA231-261.

286. S. Li and L. Huang: In vivo gene transfer via intravenous administration of cationic lipid-protamine-DNA (LPD) complexes 39, Gene Ther. 1997; 4(9):891-900.

287. N. Oku, Y. Yamazaki, M. Matsuura, M. Sugiyama, M. Hasegawa and M. Nango: A novel non-viral gene transfer system, polycation liposomes, Adv Drug Deliv Rev 2001; 52(3):209-218.

288. M. Matsuura, Y. Yamazaki, M. Sugiyama, M. Kondo, H. Ori, M. Nango and N. Oku: Polycation liposome-mediated gene transfer in vivo, Biochim Biophys Acta 2003; 1612(2):136-143.

289. S. Chono, S.D. Li, C.C. Conwell and L. Huang: An efficient and low immunostimulatory nanoparticle formulation for systemic siRNA delivery to the tumor, J Control Release 2008; 131(1):64-69.

290. A. Akinc, A. Zumbuehl, M. Goldberg, E.S. Leshchiner, V. Busini, N. Hossain, S.A. Bacallado, D.N. Nguyen, J. Fuller, R. Alvarez, A. Borodovsky, T. Borland, R. Constien, A. de Fougerolles, J.R. Dorkin, J.K. Narayanannair, M. Jayaraman, M. John, V. Koteliansky, M. Manoharan, L. Nechev, J. Qin, T. Racie, D. Raitcheva, K.G. Rajeev, D.W. Sah, J. Soutschek, I. Toudjarska, H.P. Vornlocher, T.S. Zimmermann, R. Langer and D.G. Anderson: A combinatorial library of lipid-like materials for delivery of RNAi therapeutics, Nat.Biotechnol. 2008; 26(5):561-569.

291. E. Mathew, G.E. Hardee, C.F. Bennett and K.D. Lee: Cytosolic delivery of antisense oligonucleotides by listeriolysin O-containing liposomes, Gene Ther 2003; 10(13):1105-1115.

292. J. Haensler and F.C. Szoka, Jr.: Polyamidoamine cascade polymers mediate efficient transfection of cells in culture, Bioconjug Chem 1993; 4(5):372-379.

293. E. Wagner: The Silent (R)evolution of Polymeric Nucleic Acid Therapeutics, Pharm.Res 2008; 25(12):2920-2923.

7 REFERENCES

294. A. Meister and M.E. Anderson: Glutathione, Annu.Rev.Biochem. 1983; 52711-760.
295. G. Zuber, E. Dauty, M. Nothisen, P. Belguise and J.P. Behr: Towards synthetic viruses, Adv.Drug Deliv.Rev. 2001; 52(3):245-253.
296. D.M. Dykxhoorn, C.D. Novina and P.A. Sharp: Killing the messenger: short RNAs that silence gene expression, Nat.Rev.Mol.Cell Biol. 2003; 4(6):457-467.
297. C.D. Novina and P.A. Sharp: The RNAi revolution, Nature 2004; 430(6996):161-164.
298. M. Thomas and A.M. Klibanov: Enhancing polyethylenimine's delivery of plasmid DNA into mammalian cells 1, Proc.Natl.Acad.Sci.U.S.A 2002; 99(23):14640-14645.
299. S. Nimesh, A. Aggarwal, P. Kumar, Y. Singh, K.C. Gupta and R. Chandra: Influence of acyl chain length on transfection mediated by acylated PEI nanoparticles, Int.J.Pharm. 2007; 337(1-2):265-274.
300. A. Mecke, I.J. Majoros, A.K. Patri, J.R. Baker, Jr., M.M. Holl and B.G. Orr: Lipid bilayer disruption by polycationic polymers: the roles of size and chemical functional group, Langmuir 2005; 21(23):10348-10354.
301. M. Thomas, J.J. Lu, Q. Ge, C. Zhang, J. Chen and A.M. Klibanov: Full deacylation of polyethylenimine dramatically boosts its gene delivery efficiency and specificity to mouse lung, Proc.Natl.Acad.Sci.U.S.A 2005; 102(16):5679-5684.
302. M. Lecocq, S. Wattiaux-De Coninck, N. Laurent, R. Wattiaux and M. Jadot: Uptake and intracellular fate of polyethylenimine in vivo, Biochem Biophys Res Commun 2000; 278(2):414-418.
303. Y. Takakura, T. Fujita, M. Hashida and H. Sezaki: Disposition characteristics of macromolecules in tumor-bearing mice, Pharm.Res. 1990; 7(4):339-346.
304. P. Dubruel, B. Christiaens, B. Vanloo, K. Bracke, M. Rosseneu, J. Vandekerckhove and E. Schacht: Physicochemical and biological evaluation of cationic polymethacrylates as vectors for gene delivery, Eur J Pharm Sci 2003; 18(3-4):211-220.
305. T. Inoue, M. Sugimoto, T. Sakurai, R. Saito, N. Futaki, Y. Hashimoto, Y. Honma, I. Arai and S. Nakaike: Modulation of scratching behavior by silencing an endogenous cyclooxygenase-1 gene in the skin through the administration of siRNA, J Gene Med. 2007; 9(11):994-1001.
306. P.A. Brown, A.S. Khan and R. Draghia-Akli: Delivery of DNA into skeletal muscle in large animals, Methods Mol Biol 2008; 423215-224.
307. Y. Takei, T. Nemoto, P. Mu, T. Fujishima, T. Ishimoto, Y. Hayakawa, Y. Yuzawa, S. Matsuo, T. Muramatsu and K. Kadomatsu: In vivo silencing of a molecular target by short interfering RNA electroporation: tumor vascularization correlates to delivery efficiency, Mol Cancer Ther 2008; 7(1):211-221.
308. F. Scherer, M. Anton, U. Schillinger, J. Henke, C. Bergemann, A. Kruger, B. Gansbacher and C. Plank: Magnetofection: enhancing and targeting gene delivery by magnetic force in vitro and in vivo, Gene Ther 2002; 9(2):102-109.
309. S. Huth, J. Lausier, S.W. Gersting, C. Rudolph, C. Plank, U. Welsch and J. Rosenecker: Insights into the mechanism of magnetofection using PEI-based magnetofectins for gene transfer, J.Gene Med. 2004; 6(8):923-936.
310. C.H. Lee, E.Y. Kim, K. Jeon, J.C. Tae, K.S. Lee, Y.O. Kim, M.Y. Jeong, C.W. Yun, D.K. Jeong, S.K. Cho, J.H. Kim, H.Y. Lee, K.Z. Riu, S.G. Cho and S.P. Park: Simple, efficient, and reproducible gene transfection of mouse embryonic stem cells by magnetofection, Stem Cells Dev. 2008; 17(1):133-141.

7 REFERENCES

311. C. Huttinger, J. Hirschberger, A. Jahnke, R. Kostlin, T. Brill, C. Plank, H. Kuchenhoff, S. Krieger and U. Schillinger: Neoadjuvant gene delivery of feline granulocyte-macrophage colony-stimulating factor using magnetofection for the treatment of feline fibrosarcomas: a phase I trial, J Gene Med. 2008; 10(6):655-667.

312. D.L. Miller, S.V. Pislaru and J.E. Greenleaf: Sonoporation: mechanical DNA delivery by ultrasonic cavitation, Somat.Cell Mol.Genet. 2002; 27(1-6):115-134.

313. R.K. Schlicher, H. Radhakrishna, T.P. Tolentino, R.P. Apkarian, V. Zarnitsyn and M.R. Prausnitz: Mechanism of intracellular delivery by acoustic cavitation, Ultrasound Med.Biol. 2006; 32(6):915-924.

314. D. Sheyn, N. Kimelman-Bleich, G. Pelled, Y. Zilberman, D. Gazit and Z. Gazit: Ultrasound-based nonviral gene delivery induces bone formation in vivo, Gene Ther 2008; 15(4):257-266.

315. A. Hogset, L. Prasmickaite, B.O. Engesaeter, M. Hellum, P.K. Selbo, V.M. Olsen, G.M. Maelandsmo and K. Berg: Light directed gene transfer by photochemical internalisation, Curr.Gene Ther. 2003; 3(2):89-112.

316. A. Hogset, L. Prasmickaite, P.K. Selbo, M. Hellum, B.O. Engesaeter, A. Bonsted and K. Berg: Photochemical internalisation in drug and gene delivery, Adv.Drug Deliv.Rev. 2004; 56(1):95-115.

317. J. Kloeckner, L. Prasmickaite, A. Hogset, K. Berg and E. Wagner: Photochemically enhanced gene delivery of EGF receptor-targeted DNA polyplexes, J Drug Target 2004; 12(4):205-213.

318. A. Ndoye, G. Dolivet, A. Hogset, A. Leroux, A. Fifre, P. Erbacher, K. Berg, J.P. Behr, F. Guillemin and J.L. Merlin: Eradication of p53-mutated head and neck squamous cell carcinoma xenografts using nonviral p53 gene therapy and photochemical internalization, Mol.Ther. 2006; 13(6):1156-1162.

319. A. Bonsted, E. Wagner, L. Prasmickaite, A. Hogset and K. Berg: Photochemical enhancement of DNA delivery by EGF receptor targeted polyplexes, Methods Mol Biol 2008; 434171-181.

320. C.H. Ahn, S.Y. Chae, Y.H. Bae and S.W. Kim: Biodegradable poly(ethylenimine) for plasmid DNA delivery 1, J Control Release 2002; 80(1-3):273-282.

321. W.J. Kim, C.W. Chang, M. Lee and S.W. Kim: Efficient siRNA delivery using water soluble lipopolymer for anti-angiogenic gene therapy, J Control Release 2007; 118(3):357-363.

322. M. Frank-Kamenetsky, A. Grefhorst, N.N. Anderson, T.S. Racie, B. Bramlage, A. Akinc, D. Butler, K. Charisse, R. Dorkin, Y. Fan, C. Gamba-Vitalo, P. Hadwiger, M. Jayaraman, M. John, K.N. Jayaprakash, M. Maier, L. Nechev, K.G. Rajeev, T. Read, I. Rohl, J. Soutschek, P. Tan, J. Wong, G. Wang, T. Zimmermann, A. de Fougerolles, H.P. Vornlocher, R. Langer, D.G. Anderson, M. Manoharan, V. Koteliansky, J.D. Horton and K. Fitzgerald: Therapeutic RNAi targeting PCSK9 acutely lowers plasma cholesterol in rodents and LDL cholesterol in nonhuman primates, Proc.Natl.Acad.Sci U.S.A 2008; 105(33):11915-11920.

323. A.E. Smith and A. Helenius: How viruses enter animal cells, Science 2004; 304(5668):237-242.

324. P.J. Butler, J.I. Harris, B.S. Hartley and R. Leberman: Reversible blocking of peptide amino groups by maleic anhydride, Biochem.J. 1967; 103(3):78P-79P.

325. E.S. de la and E. Palacian: Dimethylmaleic anhydride, a specific reagent for protein amino groups, Biochem.Cell Biol 1989; 67(1):63-66.

326. S.A. Moschos, S.W. Jones, M.M. Perry, A.E. Williams, J.S. Erjefalt, J.J. Turner, P.J. Barnes, B.S. Sproat, M.J. Gait and M.A. Lindsay: Lung delivery studies using siRNA conjugated to TAT(48-60) and penetratin reveal peptide induced reduction in gene expression and induction of innate immunity, Bioconjug Chem 2007; 18(5):1450-1459.

327. S. Takae, K. Miyata, M. Oba, T. Ishii, N. Nishiyama, K. Itaka, Y. Yamasaki, H. Koyama and K. Kataoka: PEG-detachable polyplex micelles based on disulfide-linked block catiomers as bioresponsive nonviral gene vectors, J Am Chem Soc 2008; 130(18):6001-6009.

328. S. Boeckle, E. Wagner, M. Ogris: C- versus N-terminally linked melittin-polyethylenimine conjugates: the site of linkage strongly influences activity of DNA polyplexes, J Gene Med. 2005 Oct; 7(10):1335-47.

8 ACKNOWLEDGEMENTS

First of all, I am very thankful to all my colleagues for their support and the great time we spent together during the last three years. Without you, work would not have attracted as much fun and this thesis would not have been possible.

Foremost, I want to express my sincere gratitude to my supervisor Prof. Dr. Ernst Wagner for giving me the opportunity to join his excellent research group and the trust and confidence placed in me. Many thanks for your professional guidance and skillful scientific support generating a creative and instructive atmosphere which was essential for the success of this thesis.

I am also very grateful to Dr. Arkadi Zintchenko and Dr. Manfred Ogris for the close collaboration between "in vitro" and "in vivo" and for the many helpful advices and discussions and especially to "Fredl" for continually supplying us with "Apfelstrudl". Many thanks to Dr. Jaroslav Pelisek for guiding me during the beginning of my PhD thesis and supporting me with his practical experience in the field of in vitro nucleic acid delivery. Also I would like to thank Dr. Martina Rüffer for her continuous support during the student courses and in cooperation with Mrs. Annemi Willmann for the help in "ploughing" me across the botany.

A big thank goes to all our technicians Anna, Markus, Melinda, Miriam, Ursula and Wolfgang for taking care of all the many things that come up in order to keep the lab routine running from cell culture to animal house. Anna, great thanks to you for always performing the many innumerable luminometer measurements of hundreds (or thousands?) of well-plates for me. Wolfgang, thank you for your immediate help in any kinds of technical problems especially when the PC intended having a crash down again.

Special thanks to my direct lab-mate Martin for having a great time in the laboratory exploring science and having lots of fun, such as with the deafening notorious "Santa Claus". Thanks to Verena, Veronika and Christian for the various polymer syntheses and Nicole, Gelja and Katarina for performing the in vivo experiments. I would like to thank Edith for the help with NMR analysis as well as Terese and Arzu for the fruitful qPCR discussions. Many thanks also to my cell culture colleagues Thomas and Daniel and all the other former and newer PhD students for all the fun we had.

A very special and heartfelt thank you goes to Kristin for her enduring patience, effort and indispensable affection all the time.

Finally, I want to thank most of all my parents for their sincere appreciation and everlasting support in all aspects of my life for so many reasons.

Die VDM Verlagsservicegesellschaft sucht für wissenschaftliche Verlage abgeschlossene und herausragende

Dissertationen, Habilitationen, Diplomarbeiten, Master Theses, Magisterarbeiten usw.

für die kostenlose Publikation als Fachbuch.

Sie verfügen über eine Arbeit, die hohen inhaltlichen und formalen Ansprüchen genügt, und haben Interesse an einer honorarvergüteten Publikation?

Dann senden Sie bitte erste Informationen über sich und Ihre Arbeit per Email an *info@vdm-vsg.de*.

Sie erhalten kurzfristig unser Feedback!

VDM Verlagsservicegesellschaft mbH
Dudweiler Landstr. 99   Telefon +49 681 3720 174
D - 66123 Saarbrücken   Fax   +49 681 3720 1749
www.vdm-vsg.de

Die VDM Verlagsservicegesellschaft mbH vertritt

Printed by Books on Demand GmbH, Norderstedt / Germany